AF691359

WITHDRAWN
IOWA STATE UNIVERSITY
LIBRARY

Guidelines for Chemical Transportation Safety, Security, and Risk Management

This book is one in a series of process safety guideline and concept books published by the Center for Chemical Process Safety (CCPS). Please go to www.wiley.com/go/ccps for a full list of titles in this series.

Guidelines for Chemical Transportation Safety, Security, and Risk Management

Center for Chemical Process Safety
New York, New York

CCPS
CENTER FOR CHEMICAL PROCESS SAFETY
An AIChE Industry Technology Alliance

WITHDRAWN
IOWA STATE UNIVERSITY
LIBRARY

WILEY
A JOHN WILEY & SONS, INC., PUBLICATION

It is sincerely hoped that the information presented in this document will lead to an even more impressive safety record for the entire industry. However, neither the American Institute of Chemical Engineers, its consultants, CCPS Technical Steering Committee and Subcommittee members, their employers, their employers' officers and directors, nor AcuTech Consulting Group, Inc. and its employees warrant or represent, expressly or by implication, the correctness or accuracy of the content of the information presented in this document. As between (1) American Institute of Chemical Engineers, its consultants, CCPS Technical Steering Committee and Subcommittee members, their employers, their employers' officers and directors, and AcuTech Consulting Group, Inc. and its employees and (2) the user of this document, the user accepts any legal liability or responsibility whatsoever for the consequence of its use or misuse.

Copyright © 2008 by American Institute of Chemical Engineers, Inc. All rights reserved.

A Joint Publication of the Center for Chemical Process Safety of the American Institute of Chemical Engineers and John Wiley & Sons, Inc.

Published by John Wiley & Sons, Inc., Hoboken, New Jersey.
Published simultaneously in Canada.

No part of this publication may be reproduced, stored in a retrieval system, or transmitted in any form or by any means, electronic, mechanical, photocopying, recording, scanning, or otherwise, except as permitted under Section 107 or 108 of the 1976 United States Copyright Act, without either the prior written permission of the Publisher, or authorization through payment of the appropriate per-copy fee to the Copyright Clearance Center, Inc., 222 Rosewood Drive, Danvers, MA 01923, (978) 750-8400, fax (978) 750-4470, or on the web at www.copyright.com. Requests to the Publisher for permission should be addressed to the Permissions Department, John Wiley & Sons, Inc., 111 River Street, Hoboken, NJ 07030, (201) 748-6011, fax (201) 748-6008, or online at http://www.wiley.com/go/permission.

Limit of Liability/Disclaimer of Warranty: While the publisher and author have used their best efforts in preparing this book, they make no representations or warranties with respect to the accuracy or completeness of the contents of this book and specifically disclaim any implied warranties of merchantability or fitness for a particular purpose. No warranty may be created or extended by sales representatives or written sales materials. The advice and strategies contained herein may not be suitable for your situation. You should consult with a professional where appropriate. Neither the publisher nor author shall be liable for any loss of profit or any other commercial damages, including but not limited to special, incidental, consequential, or other damages.

For general information on our other products and services or for technical support, please contact our Customer Care Department within the United States at (800) 762-2974, outside the United States at (317) 572-3993 or fax (317) 572-4002.

Wiley also publishes its books in a variety of electronic formats. Some content that appears in print may not be available in electronic format. For information about Wiley products, visit our web site at www.wiley.com.

Library of Congress Cataloging-in-Publication Data is available.

ISBN 978-0471-78242-1

Printed in the United States of America.

10 9 8 7 6 5 4 3 2 1

CONTENTS

Preface . . . xi
Acknowledgments . . . xiii
Items on the CD . . . xv
Glossary . . . xix

1 INTRODUCTION 1

1.1 Key Stakeholders in the Supply Chain and Risk Management Process 2
1.2 Transportation Risk Management 3
1.3 Using These Guidelines 7

2 PRIMARY MANAGEMENT SYSTEMS 9

2.1 Regulatory Compliance 10
2.1.1 UN Model Framework 10
2.1.2 International Regulations 11
2.1.3 Regional Regulations 13
2.1.4 Harmonization of Transportation Regulations 15
2.2 Essential Components of a Transportation Management System 16
2.2.1 Industry Standards and Guidelines 18
2.2.2 Operational Management 18
2.2.3 Emergency Preparedness and Response 19
2.2.4 Incident Reporting and Investigation 20
2.2.5 Management of Change 21
2.2.6 Transportation Risk Management System Auditing 22
2.3 XYZ Chemical Example—Primary Management Systems 23

3 RISK ASSESSMENT FUNDAMENTALS 25

3.1 Safety Risk Assessment Concepts 25
3.2 Risk Definitions 27
3.2.1 Hazards 28
3.2.2 Initiating Events 28
3.2.3 Incident Consequences 29
3.2.4 Incident Impacts 31

3.2.5 Likelihood 33
3.2.6 Differences in Safety and Security Risk Definitions 33
3.3 Risk Analysis Protocol 34
3.4 Identification and Prioritization Activities 36
3.4.1 Cataloging Hazardous Materials and Modes of Transportation 36
3.4.2 Identifying Sensitive Areas Along the Transportation Route 37
3.4.3 Interactions with Other Stakeholders in the Supply Chain 37
3.4.4 Identification and Prioritization Summary 38
3.5 XYZ Chemical Example—Identification and Prioritization 39

4 QUALITATIVE AND SEMI-QUANTITATIVE RISK ANALYSIS 43

4.1 Overview of Qualitative and Semi-Quantitative Risk Assessments 44
4.2 Qualitative Risk Analysis 46
4.2.1 Establishing Benchmarking Comparisons 46
4.2.2 Identifying Issues for Further Evaluation 48
4.2.3 Understanding the Impact of Anticipated Changes 50
4.2.4 XYZ Chemical Example—Qualitative Risk Analysis 50
4.3 Semi-Quantitative Risk Analysis 54
4.3.1 Refinement of Consequence and Likelihood Estimates and Preliminary Mitigation Considerations 57
4.3.2 Semi-Quantitative Risk Analysis Techniques 61
4.3.3 XYZ Chemical Example—Semi-Quantitative Risk Analysis 65

5 QUANTITATIVE RISK ANALYSIS 71

5.1 Overview 72
5.1.1 QRA Study Design 73
5.1.2 QRA Protocol 74
5.2 QRA Data Sources 75
5.2.1 Data for Frequency Analysis 77
5.2.2 Data for Consequence Analysis 80
5.2.3 Other QRA Data Categories 84
5.3 Presentation of Quantitative Results 85
5.3.1 QRA Results 85

5.4 XYZ Chemical Example—Quantitative Risk Analysis 92
5.4.1 Scope Definition 92
5.4.2 Data Collection 94
5.4.3 Freqeuncy Analysis 96
5.4.4 Consequence Analysis 96
5.4.5 Risk Estimation and Discussion of Results 97

6 TRANSPORTATION SECURITY CONSIDERATIONS 101

6.1 Overview of Transportation Security 101
6.2 Transportation Security Concepts 103
6.2.1 Security Risk Definitions 103
6.2.2 Security Regulations and Industry Practices 106
6.2.3 Safety and Security Synergies and Tradeoffs 111
6.3 Security Prioritization Process 111
6.4 Transportation Security Vulnerability Assessment 115
6.4.1 Consequence 115
6.4.2 Vulnerability 115
6.4.3 Threat 116
6.4.4 TSVA Methodology 117
6.4.5 TSVA Resources 119
6.5 Practical Transportation Security Elements 119
6.5.1 General Security Elements and Security Plans 120
6.5.2 Additional Considerations 123
6.6 XYZ Chemical Example—Security Analysis 124

7 RISK REDUCTION STRATEGIES 133

7.1 Risk Reduction Initiatives 133
7.1.1 Documentation and Communication of Risk Reduction Strategies 134
7.1.2 Risk Reduction Opportunities 135
7.1.3 Balancing Safety and Security 136
7.2 Factors Influencing Risk Reduction Options 138
7.2.1 Overall Supply Chain Risk Reduction Options 138
7.2.2 Pre-shipment Risk Reduction Options 140
7.2.3 Emergency Response and Post-Incident Risk Reduction Opportunities 144
7.3 Selection of Risk Reduction Options 145
7.4 XYZ Chemical Example—Risk Reduction Strategies 148
7.4.1 Qualitative Risk Analysis Reduction Options 150

7.4.2 Semi-Quantitative Risk Analysis Reduction Options 150
7.4.3 Quantitative Risk Analysis Reduction Options 152
7.4.4 Risk Reduction Options for Security 154

8 PROGRAM SUSTAINABILITY 157

8.1 Ongoing Commitment to Risk Management 158
8.2 Continuous Improvement 159
8.3 Emerging Safety and Security Trends 160
8.4 Evolving Transportation Risk Analysis Practices 161
8.5 XYZ Chemical Example—Program Sustainability 161

INDEX 163

PREFACE

The transportation of chemicals and other hazardous materials on roads, railways, waterways, through pipelines, and by other modes and systems provides vital support for today's economy and standard of living. Global trade has resulted in more transborder shipments of raw materials, consumer products, and hazardous wastes than ever before. The volume of traffic and the speed with which it moves continue to increase in both developed and developing countries, with some transportation systems operating at full capacity, thereby challenging the supporting infrastructure. Urban sprawl and population growth have brought people into closer contact with established transportation corridors, with much of the physical transportation infrastructure carrying both people and industrial shipments.

Because the release of a chemical during transit poses a potential risk to surrounding communities, property, and the environment, the transportation of hazardous materials must be carefully managed. It is widely recognized that chemical transportation is more than just a service function that fills a business void; it is a comprehensive system of activities that involves numerous stakeholders with varying backgrounds and interests. Therefore, chemical transportation should address the entire supply chain of a commodity, spanning the delivery and handling of raw materials through the offloading and unpacking of finished products. Additionally, beyond the desire for accident-free chemical transport, the evolving threat of terrorism and other security concerns should be considered as part of a transportation risk management program. Since there may be tradeoffs between safety and security, this balance can challenge even the best business plans and risk management programs.

In 1995, the Center for Chemical Process Safety (CCPS) published a book titled Guidelines for Chemical Transportation Risk Analysis. This publication provided a technical review of risk analysis techniques used to evaluate chemical transportation operations. In particular, it provided detailed methods and techniques for designing and conducting a formal transportation risk analysis, extending the risk analysis concepts developed and applied for fixed chemical process sites. Supporting the technical background and methodologies, the book included detailed examples illustrating the application of risk analysis techniques in a chemical transportation setting, and an extensive set of appendices containing case studies, data sources, and references.

In the decade since this first book was published, it has received widespread recognition as an excellent resource on transportation risk analysis. It has been used extensively by major chemical companies and shipping organizations both domestically and internationally. In order to best serve the needs and interests of its clients and stakeholders, CCPS periodically reviews the use and effectiveness of its publications and issues updates or revisions. This new book is not meant to replace the earlier 1995 edition, but rather to augment it. The new publication deals with transportation, security, and risk management on a broader basis and provides tools and methods to benefit a wider range of transportation professionals and associated stakeholders. In particular, it introduces more qualitative and practical techniques for screening, identifying, and managing higher-level risks issues that balance both safety and security. These two books should be used in combination to effectively manage and analyze transportation risk. For this reason, the original Guidelines for Chemical Transportation Risk Analysis is reproduced on the enclosed CD, as Appendix A.

This new publication provides a framework and common practices that will allow shippers and supply stakeholders to manage chemical transportation networks. The framework should embrace and utilize a company's existing systems, but must meet regulatory requirements in areas where shipping activities occur. It is written from an international perspective, recognizing that many companies transport materials across country borders and around the globe. Local customs, regulations, and working cultures may vary, but the common principles of safety, security, and risk management should always be recognized.

The development of this newly expanded transportation risk management guidelines book involved several discussions with stakeholder groups in the manufacturing, shipping, and regulatory sectors. Many participants in these sessions identified the need for a practical set of tools that would mesh with existing management practices and make optimum use of existing resources. To that end, this book attempts to provide useful tools and information covering a wide range of chemical transportation applications for all user groups. This book is a tribute to the many dedicated transportation, safety, security, and risk professionals who provided countless hours discussing issues, sharing information, and reviewing documentation to make possible both the original book and this new valuable publication.

The final chapter of this book, Program Sustainability, discusses keeping risk management practices current with changing trends and regulations. To support this need, a comprehensive set of appendices, in CD format, is included. In addition to the original 1995 book, the appendices include tools shared by stakeholder companies, links to current and pending regulations and to additional resources current at the time of publication, and the complete example of the transportation concepts that are developed throughout each chapter of this book.

ACKNOWLEDGMENTS

The American Institute of Chemical Engineers (AIChE) and the Center for Chemical Process Safety (CCPS) express their appreciation and gratitude to all members of the Chemical Transportation Risk and Security Subcommittee and their CCPS member companies for their generous support and technical contributions in the preparation of these guidelines. We would also like to acknowledge the willingness of chemical manufacturers, shipping companies, and regulatory agencies to share their risk evaluation tools and protocols, some of which are included as examples in the enclosed CD appendices. Finally, the need for this guideline was identified by Henry Ward of the Dow Chemical Company, who helped steer the initial efforts of the subcommittee and established an effective stakeholder network.

Chemical Transportation Risk and Security Subcommittee Members:

Cheryl (Cherry) Burke (Co-chair)
The Dow Chemical Company

Lou Castiglioni
Arkema Inc.

Mark Connolly
Akzo Nobel Chemicals

Brian R. Dunbobbin (Co-chair)
Air Products and Chemicals Inc.

Donna Edminster
Rhodia Inc.

Michael R. Green
BP America, Inc.

Bob Hollenbeck
3M Company (retired)

Gary Staton
DuPont Company

Stephen F. Urschel
SABIC Innovative Plastics, LLC

William Ed Waley
LyondellBasell Industries

CCPS Staff Consultant: Brian Kelly
CCPS Emeritus

Principal Author: Brad A. Fuller
AcuTech Consulting Group

The authors wish to thank the following AcuTech Consulting Group personnel for their technical contributions and review: Tara Aboyoun, Steven M. Fruchtman, Dale A. Kers, H.M. Leith, Kevin F. Molloy, David A. Moore, Martin R. Rose, and Lee Salamone.

Before publication, all CCPS books are subjected to a thorough peer review process. CCPS gratefully acknowledges the thoughtful comments and suggestions of the peer reviewers, whose work enhanced the accuracy and clarity of these guidelines.

CTRS Peer Reviewers:

Lisa Bendixen
ICF Consultants

Ron Christy
Celanese Chemicals

Graham Creedy
Canadian Chemical Producers' Association

David Gessford
Dow Chemical Company

Jean-Pierre Durand
Olin Corporation

Ronald de Nooijer
SABIC Innovative Plastics

Zhigang Qin
SABIC Innovative Plastics

Doug Reeves
Department of Transportation (retired)

Bill Rhyne
Consultant (retired)

Mark Stehly
BNSF Railway

Henry Ward
Dow Chemical Company

ITEMS ON THE CD ACCOMPANYING THIS BOOK

Appendix A: Guidelines for Chemical Transportation Risk Analysis (CCPS, 1995)

Appendix B: Industry Tools

Appendix C: Regulation Links

Appendix D: Resource Links

Appendix E: Worked TRM Example

ACRONYMS

AAR	Association of American Railroads
ACC	American Chemistry Council
ADR	International Carriage of Dangerous Goods by Road (ADR stands for Accord européen sur le transport des marchandises dangereuses par route)
AEO	Authorized Economic Operators
AIChE	American Institute of Chemical Engineers
ADN	International Carriage of Dangerous Goods by Inland Waterways
API	American Petroleum Institute
A_T	Target Attractiveness
BLEVE	Boiling Liquid Expanding Vapor Explosion
C	Consequence
CANUTEC	Canadian Transport Emergency Center
CBP	Customs and Border Protection
CCPS	Center for Chemical Process Safety
CD	Compact Disc
CFATS	Chemical Facility Anti-terrorism Standards
CFR	Code of Federal Regulations
CHEMTREC	Chemical Transportation Emergency Center
CPQRA	Chemical Process Quantitative Risk Analysis
C-TPAT	Customs-Trade Partnership Against Terrorism
DGR	Dangerous Goods Regulations
DHS	U.S. Department of Homeland Security
DOT	U.S. Department of Transportation
EHS	Environment, Health, and Safety
EU	European Union
FAA	Federal Aviation Administration
FAR	Fatal Accident Rate
FAST	Free and Secure Trade
FMCSA	Federal Motor Carrier Safety Administration
F-N	Frequency-Number
FRA	Federal Railway Administration
GHS	Globally Harmonized System
GPS	Global Positioning System

HAZMAT	Hazardous Material
HCl	Hydrochloric Acid
HMTA	Hazardous Materials Transportation Act
HSAS	Homeland Security Advisory System
IATA	International Air Transport Association
ICAO	International Civil Aviation Organization
IMDG	International Maritime Dangerous Goods Code
IMO	International Maritime Organization
ISO	International Organization for Standardization
ISPS	International Ship and Port Facility Security
IST	Inherently Safer Technology
IT	Information Technology
L_A	Likelihood of Adversary Attack
L_{AS}	Likelihood of Adversary Success
LSP	Logistic Service Provider
MARSEC	Maritime Security
MTSA	Maritime Transportation Security Act
NACD	National Association of Chemical Distributors
NOMs	Normas Oficiales Mexicanas (Official Mexican Standards)
NOV	Notice of Violation
NPRA	National Petroleum Refiners Association
NVIC	Navigation and Vessel Inspection Circular
OITF	Intergovernmental Organization for International Carriage by Rail
PHA	Process Hazards Analysis
PHMSA	Pipeline and Hazardous Materials Safety Administration
PPM	Parts Per Million
PSI	Pounds Per Square Inch
QA	Quality Assurance
QC	Quality Control
QRA	Quantitative Risk Analysis
RDP	Responsible Distribution Process
RFID	Radio-Frequency Identification
RID	Transport of Dangerous Goods by Rail (RID stands for Reglement International concernant le transport des marchandises Dangereuses par chemin de fer)
SAFE	Framework of Standards to Secure and Facilitate Global Trade
SCT	Mexican Secretariat for Communications and Transport
SOCMA	Synthetic Organic Chemical Manufacturers Association
SOLAS	Safety of Life at Sea
SVA	Security Vulnerability Analysis
T	Threat

TDG	Transportation of Dangerous Goods Act
TNO	Netherlands Organization for Applied Scientific Research
TRM	Transportation Risk Management
TSA	Transportation Security Administration
TSVA	Transportation Security Vulnerability Analysis
TWIC	Transportation Worker Identification Credential
UN	United Nations
UNECE	United Nations Economic Commission for Europe
USCG	United States Coast Guard
V	Vulnerability
VAM-CF	Vulnerability Assessment Methodology for Chemical Facilities
VCE	Vapor Cloud Explosion
WCO	World Customs Organization

GLOSSARY

Accident-Initiated Event An event (or the first event in an event sequence) that is caused by a movement-related transportation accident, such as a train derailment or a barge grounding.

Acute Risk A risk arising from a short-term event such as a release causing a fire, explosion, or short-duration toxic exposure.

Adversary Any individual, group, organization, or government that conducts activities, or has the intention and capability to conduct activities, detrimental to critical assets. An adversary can include intelligence services of host nations, or third-party nations, political and terrorist groups, criminals, disgruntled employees or contractors, and private interests. Adversaries also can include insiders, outsiders, or the two acting in collusion.

Alert Levels Describes a progressive, qualitative measure of the likelihood of terrorist actions, from negligible to imminent, based on government or company intelligence information. Different security measures may be implemented at each alert level based on the level of threat.

Allision The act of a moving vessel striking against, or colliding with, a stationary object.

Asset Any person, environment, facility, material, information, business reputation, or activity that has positive value to an owner. The asset may have value to an adversary, as well as an owner, although the nature and magnitude of those values may differ.

Asset Category Assets may be categorized in many ways, including people, chemicals, information, environment, equipment, facilities, and activities/operations.

Atmospheric Stability A measure of the degree of atmospheric turbulence commonly defined in terms of the vertical temperature gradient. These are designated by the letters A through G.

Average Individual Risk Three average individual risks are used in this book:

1. Exposed population is the individual risk averaged over the population that is exposed to a particular hazard.
2. Total population is the individual risk averaged over a predetermined population, without regard to whether or not all people in that population are actually exposed to the risk.
3. Exposed hours/worked hours may be calculated for the duration of the activity or may be averaged over the working day.

Audit A systematic, independent review to verify conformance with prescribed standards of care using a well-defined review process to ensure consistency and to allow the auditor to reach defensible conclusions.

Basic Event A fault tree event is sufficiently basic if no further development is necessary (e.g., equipment item failure, human failure, or external event).

Benchmarking The comparison of current operating practices to internal or external company practices, industry best practices, and regulatory standards.

Benefit Amount of expected risk reduction.

BLEVE A boiling liquid expanding vapor explosion, which occurs from the sudden release of a large mass of pressurized liquid to the atmosphere. A primary cause is an external flame impinging on the shell of a vessel above the liquid level, weakening the shell and resulting in sudden rupture.

Bounding Group (of Incidents) A small number of incidents selected to bracket the spectrum of possible incidents, which may include those catastrophic incidents sometimes referred to as the worst credible incident and the worst possible incident.

Capability When assessing the capability of an adversary, two distinct capabilities need to be considered. The first is the capability to obtain, damage, or destroy the asset. The second is the adversary's capability to use the asset to achieve its objectives once the asset is obtained, damaged, or destroyed.

Chronic Risk A risk arising from a long-term persistent exposure.

Collision An incident involving two or more moving vessels that collide while meeting, overtaking, or crossing.

Commodity Data Information about the hazardous material(s) that is being shipped such as physical properties of the chemical, packaging types (including quantity in transit and storage conditions), annual shipments, time of day the shipment moves across the route or segments of the route, and breakout of loaded versus empty miles.

Conditional Probability Probability of occurrence given that a precursor event has occurred.

Consequence The direct, undesirable result of an incident outcome, specifically, the impacts resulting from a release of a hazardous material. Consequence is generally a function of the hazards of the material released, the extent of the release, and the presence of receptors (people, ecosystems, property, etc.).

Consequence Analysis The analysis of the expected effects of incident outcome cases independent of frequency or probability.

Continuous Improvement Through commitment, self-evaluation, and the flexibility to change, improvements in risk management results and efficiency should be sought continuously.

Cost Includes tangible items such as money and equipment as well as the operational costs associated with the implementation of risk reduction options. There are also intangible costs such as loss of productivity, moral considerations, political embarrassment, and a variety of others. Costs may be borne by the individuals who are affected or the corporations they work for, or they may involve macroeconomic costs to society.

Cost-Benefit Analysis Part of the management decision-making process in which the costs and benefits of each risk reduction option are compared and the most appropriate alternative is selected.

Countermeasure An action taken or a physical capability provided whose principal purpose is to reduce or eliminate one or more vulnerabilities.

CPQRA A chemical process quantitative risk analysis is the process of hazard identification followed by numerical evaluation of incident consequences and frequencies, and their combination into an overall measure of risk when applied to the chemical process industry. It is particularly applicable to episodic events. It differs from, but is related to, a probabilistic risk assessment (PRA), a quantitative tool used in the nuclear industry.

Critical Infrastructure Systems and assets, whether physical or virtual, so vital that the loss, interruption, incapacity, or destruction of which (1) would have a negative or debilitating effect on the security, economic security, public health, or safety of a nation, region, or any local government, or (2) cause national or regional catastrophic effects.

Delay A countermeasure strategy that is intended to provide various barriers to slow the progress of an adversary so as to prevent an attack or theft, increase the time necessary to cause an event, or assist in apprehension of an adversary and prevention of theft.

Demographic Data The description of the population along the transportation route, which can include the development and use of representative urban and rural population densities along the route or more precise estimates (e.g., census data).

Detection A countermeasure strategy intended to identify an adversary attempting to plan or commit a chemical security event or other criminal activity via either real-time observations or pre-incident analysis of the activities/intelligence.

Deterrence A countermeasure strategy intended to prevent or discourage the occurrence of a breach of security by means of fear or doubt.

Effect Models Models that predict effects of incident outcomes, usually with respect to human injury, fatality, or property damage.

Event An occurrence involving equipment performance or human action, or an occurrence external to the system that causes a system upset. In this book, an event is associated with an incident either as the cause or a contributing cause of the incident or as a response to the initiating event.

Event Sequence A specific unplanned sequence of events composed of initiating events and intermediate events that may lead to an incident.

Event Tree (Analysis) A graphical logic model that identifies and quantifies possible outcomes following an initiating event.

Explosions A release of energy that causes a pressure discontinuity or blast wave.

Fireball The atmospheric burning of a fuel-air cloud in which the energy is mostly emitted in the form of radiant heat. The inner core of the fuel release consists of almost pure fuel whereas the outer layer in which ignition first occurs is a flammable fuel-air mixture. As buoyancy forces of the hot gases begin to dominate, the burning cloud rises and becomes more spherical in shape.

Flash Fire The combustion of a flammable vapor and air mixture in which flame passes through that mixture at less than sonic velocity, such that negligible damaging overpressure is generated.

F-N Curve A plot of cumulative frequency versus consequences (often expressed as number of fatalities).

Frequency Number of occurrences of an event per unit of time.

Frequency Data The data required to generate accident and non-accident rates, the probability of a release following an accident, the range of release sizes to be considered, and the probabilities of various outcomes of release (i.e., toxic, flammable, explosive).

Generic Data Data that are built using inputs from all operations within a company, from literature sources, from past TRA reports, and from commercial databases such as accident data for all vehicles vs. trucks only vs. certain types of trucks.

Hazard Inherent property or characteristic of a material, system, or process that has the potential for causing serious injury to people and/or property or environmental damage.

Hazard Zone For an incident that produces an outcome such as toxic release, the hazard zone is the area over which the airborne concentration equals or exceeds some level of concern. For a flammable release, the area of effect is based on a specified level of thermal radiation. For a release that results in explosion, this is the area defined by specified overpressure levels.

Historical Incident Data Data collected and recorded from past incidents.

Identification and Prioritization Consists of (1) cataloging the hazardous materials and modes of transportation, (2) identifying sensitive areas and potential points of failure along the transit route, and (3) understanding interactions with other stakeholders in the supply chain. This process enables the identification of shipments that may require special attention, including escalating issues/scenarios for more detailed risk analysis.

Incident The loss of containment of material or energy.

Incident Outcome The physical manifestation of the incident: for toxic materials, the incident outcome is a toxic release, while for flammable materials, the incident outcome could be a boiling liquid expanding vapor explosion (BLEVE), flash fire, vapor cloud explosion (VCE), etc. For example, the incident outcome for a leak of chlorine from a railcar is a toxic release.

Individual Risk The risk to a person in the vicinity of a hazard. This includes the nature of the injury to the individual, the likelihood of the injury occurring, and the time period over which the injury might occur.

Initiating Event The first event in an event sequence.

Intent A course of action an adversary intends to follow. When assessing threats, security professionals need to evaluate intent as well as capabilities. To determine the intent and motivation of an adversary, an adversary's goals and objectives, as well as specific events that might trigger an adversary's act, must be closely examined. The questions that should be asked about intent are: "Does the adversary have a current or projected need for this asset? Does the adversary seek to deny or destroy the use of this asset?"

Jet Fire Type of fire resulting from pressurized release of gas and/or liquid.

Layers of Protection A concept whereby several independent devices, systems, or actions are provided to reduce the likelihood and severity of an undesired event.

Likelihood A measure of the expected probability or frequency of occurrence of an event such as events per year.

Likelihood of Adversary Success (L_{AS}) The potential for causing a catastrophic event by defeating the countermeasures. L_{AS} is an estimate of whether the security countermeasures will thwart or withstand the attempted attack or whether the attack will circumvent or exceed the existing security measures. This measure represents a surrogate for the conditional probability of success of the event.

Management System A formally established set of activities designed to produce specific results in a consistent manner on a sustainable basis.

Mode of Transportation The specific type of transportation of concern (for example, truck, rail, barge, pipeline).

Non-Accident-Initiated Event An event due to causes unassociated with the movement-related aspects of transportation such as improperly closed valves or pressure build-up due to contamination.

Physical Models Models that provide quantitative information on source rates and extent of damage (thermal radiation, explosion overpressure, or concentration of dispersing vapor clouds).

Pool Fire The combustion of material evaporating from a layer of liquid at the base of the fire.

Prevention The process of eliminating or preventing the hazards or risks associated with a particular activity. Prevention is sometimes used to describe actions taken in advance to reduce the likelihood of an undesired event.

Primary Management System The primary management elements for transportation of hazardous materials, including regulatory compliance, standards and guidelines, operational management, emergency preparedness and response, incident reporting, management of change, and auditing.

Probability The expression for the likelihood of occurrence of an event or an event sequence during an interval of time or the likelihood of the success or failure of an event on test or demand. By definition, probability must be expressed as a number ranging from 0 to 1.

Program Sustainability An ongoing commitment to managing risk, improving continuously, reacting to emerging trends, and keeping current on evolving transportation risk analysis practices.

Qualitative Risk Analysis Based primarily on description and comparison using historical experience and engineering judgment, with little quantification of the hazards, consequences, likelihood, or level of risk.

Quantitative Risk Analysis (QRA) The development of a quantitative estimate of risk based on engineering evaluation and mathematical techniques for combining estimates of incident consequence and frequencies.

Response A security strategy to neutralize the adversary or to evacuate, shelter in place, call local authorities, control a release, or take other mitigation actions.

Risk A measure of human injury, environmental damage, or economic loss in terms of both the incident likelihood and the magnitude of the loss or injury.

Risk Analysis The process (qualitative or quantitative) of evaluating consequence and likelihood and estimating risk.

Risk Assessment The process by which the results of a risk analysis (for example, risk estimates) are used to make decisions, either through relative ranking of risk reduction strategies or through comparison with risk targets.

Risk Contour Lines that connect points of equal individual risk around the facility ("iso-risk" lines).

Risk Estimation Combining the estimated consequences and likelihood of all incident outcomes from all selected incidents to provide a measure of risk.

Risk Evaluation Comparison of results against evaluation criteria used for making decisions regarding the need for risk mitigation.

Risk Evaluation Criteria A qualitative or quantitative expression of the level of risk that an individual or organization is willing to assume in return for the benefits obtained from the associated activity.

Risk Management The systematic application of management policies, procedures, and practices to the tasks of analyzing, assessing, and controlling risk in order to protect employees, the general public, and the environment as well as company assets while avoiding business interruptions.

Risk Measures Ways of combining information on likelihood with the magnitude of loss or injury such as risk indices, individual risk measures, and societal risk measures.

Risk Reduction Development, comparison, and selection of options to reduce risk to a target level, if needed, or as needed.

Route Data Information regarding the route, including conditions of the infrastructure (such as road and rail condition and speed restrictions), specific segments of the route identified for analysis (including meteorological conditions), and the location of sensitive receptors along the route (such as urban and rural population densities, environmental crossings).

Security Plan A document that describes a plan to address security issues and related events including security assessment and mitigation options. This includes security alert levels and response measures to security threats.

Security Risk The potential for damage to, or loss of, an asset. Risk, in the context of chemical process security, is the potential for the intentional event outcome to be realized. Typical examples include an intentional release of hazardous materials from containment, the theft of chemicals that could later be used as weapons, the contamination of chemicals that may later harm the public, the economic costs of the damage, or disruption of the chemical process or other nearby critical infrastructure. Therefore, risk is an expression of the likelihood (L_{AS}) that a specific vulnerability (V) of a particular attractive target (A_T) will be exploited by a defined threat (T) to cause a given consequence (C).

Security Vulnerability Analysis (SVA) The process of determining the likelihood of an adversary successfully exploiting a vulnerability and the resulting degree of damage or impact. A SVA is not a quantitative risk analysis, but is performed qualitatively using a structured and repeatable method and the best judgment of safety, security, and transportation personnel. A qualitative determination of risk is the desired outcome of a SVA in order to provide the basis for ranking the security-related risks and thus to establish priorities for the application of countermeasures.

Semi-Quantitative Risk Analysis Risk analysis methodology that includes some degree of quantification of consequence, likelihood, and/or risk level.

Sensitivity The sensitivity of a measure to a parameter is defined as the change in the measure per unit change in that parameter.

Societal Risk A measure of risk to a group of people. It is most often expressed in terms of the frequency distribution of multiple casualty events.

Target Attractiveness An estimate of the value of a target to an adversary based on the following factors:

- Potential for mass casualties/fatalities
- Extensive property damage
- Proximity to national assets or landmarks
- Possible disruption or damage to critical infrastructure
- Disruption of the national, regional, or local economy
- Ease of access to target
- Media attention or possible interest of the media
- Company reputation and brand exposure

Experience has shown that, particularly for terrorist attacks, certain targets better accomplish the objectives of the adversaries than others. Consideration should be given to these factors in defining the threat and in determining the need for any enhanced countermeasures.

Terrorism No one definition of terrorism has gained universal acceptance. Terrorism is defined in the Homeland Security Act of 2002 as any activity that "(A) involves an act that (i) is dangerous to human life or potentially destructive of critical infrastructure or key resources; and (ii) is a violation of the criminal laws of the United States or of any State or other subdivision of the United States; and (B) appears to be intended (i) to intimidate or coerce a civilian population; (ii) to influence the policy of a government by intimidation or coercion; or (iii) to affect the conduct of a government by mass destruction, assassination, or kidnapping."

Threat Any indication, circumstance, or event with the potential to cause the loss of, or damage to, an asset. Threat can also be defined as the intention and capability of an adversary to undertake actions that would be detrimental to critical assets.

Transect A risk measure illustrating changes in individual risk with distance from the source of the risk such as a transportation route.

Transloading Operations The act of transferring material from one transport container to another.

Transportation Security Vulnerability Analysis (TSVA) A modified SVA process specifically designed to account for specific threats, consequences, vulnerabilities, risks, and security countermeasures of transportation activities including hazardous materials in transit.

Transportation Supply Chain Beginning with the raw material supplier, through the manufacturing and distribution process, to the final end-use customer, and any residuals management, including the transportation network in between.

Uncertainty A measure, often quantitative, of the degree of doubt or lack of certainty associated with an estimate of the true value of a parameter.

Vapor Cloud Explosion (VCE) Explosive oxidation of a vapor cloud in a non-confined space (for example, not in vessels or buildings). Vapor cloud explosions in densely packed areas (within a plant, in a built-up city, etc.) may show accelerations in flame speeds and greater intensity blasts.

Vulnerability Any weakness that can be exploited by an adversary to gain access to an asset.

Worst Credible Incident The most severe plausible or reasonably believable incident, considering only incident outcomes and their consequences, of all identified incidents and their outcomes.

1
INTRODUCTION

The transportation of chemicals is necessary for the manufacturing and distribution of products within and across regional and international borders. Many of these shipments are regulated for transportation and are typically referred to as "dangerous goods" or "hazardous materials." These daily shipments via land, sea, and air are critical to economies across the globe, but these activities may also pose a potential risk to public safety and the environment if an accidental or intentional release were to occur during transit.

This book provides a comprehensive framework for managing transportation risks, addressing both safety and security practices. A network of stakeholders from industry and government were engaged in the development of this guideline so it could present currently recognized safety, security, and risk management practices. It is not intended to be an industry standard, but rather to advance the state of the art in chemical transportation safety, security, and risk management strategies, and to provide guidance for the industry.

It is not the intent of the Center for Chemical Process Safety (CCPS) to replace the original *Guidelines for Chemical Transportation Risk Assessment* (CCPS, 1995). The purpose of this book is to build upon and supplement the concepts in the original guidelines, targeting them to the broader range of individuals who are responsible for, or involved in, decisions that impact transportation safety and security on a daily basis. While the original guidelines should continue to be the industry reference for risk specialists conducting detailed quantitative risk assessments, the intent of this book, *Guidelines for Chemical Transportation Safety, Security, and Risk Management*, is to:

- Provide a resource for shippers, carriers, and others involved or interested in the management of chemical transportation risks
- Provide common terminology and understanding of the basic concepts
- Identify the functional areas and supply chain stakeholders who should be involved in transportation risk management activities
- Present a multitiered framework for managing chemical transportation risk and unifying the risk management efforts across the supply chain

- Provide useful tools and examples for the diverse range of transportation professionals
- Supplement the original CCPS transportation guidelines book by identifying new or improved methodologies and data sources
- Develop strategies that may be applied to a wide range of hazardous material and transportation situations
- Integrate safety, security, and logistics into a reference guide for making risk-based decisions

1.1 KEY STAKEHOLDERS IN THE SUPPLY CHAIN AND RISK MANAGEMENT PROCESS

Due to the complexity of a transportation supply chain, as shown in Figure 1.1, risk management is a shared responsibility of all stakeholders. The transportation supply chain should be considered as a single, complete system beginning with the raw material supplier, through the manufacturing and distribution process, to the final end-use customer, and any residuals management, including the transportation network in between. While the roles and responsibilities may differ for each stakeholder, and differ within the functional areas of an individual company, these groups need to understand how their activities and actions can impact the risk to the overall supply chain. The goal of this guideline is to assist in the development of transportation risk management (TRM) programs where information from these various groups can together support decision-making that balances risk with operational efficiencies, while optimizing the entire supply chain's activities.

Given the complexities of the chemical industry, the intended audience for this guideline is extensive and not limited to just shippers. To manage the risk across the entire supply chain, the following groups can benefit from the primary management systems, risk concepts, risk assessment tools and techniques, examples, and overarching risk management framework presented in this guideline:

- Logistics service providers and managers
- Shippers
- Chemical manufacturing companies
- Chemical distribution companies
- Carriers
- Business unit managers and owners
- Transportation managers
- Safety professionals
- Security professionals
- Risk professionals
- Regional and global government regulators
- Insurers
- Industry associations

Figure 1.1 Transportation Supply Chain

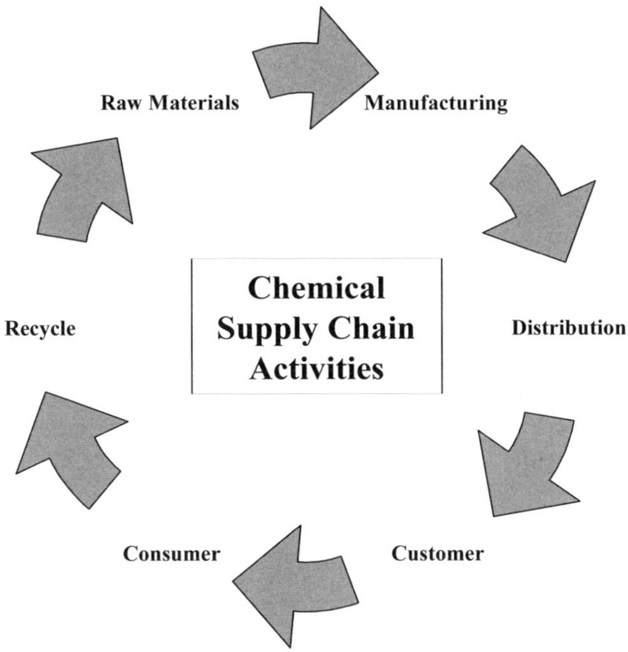

1.2 TRANSPORTATION RISK MANAGEMENT

A major goal of the safe and secure transportation of hazardous materials is a reduction in incidents that could lead to a release or to misuse. To date, the achievements in safety are the result of regulations, industry standards, individual company initiatives, and emergency response preparedness, as well as investments in training, systems, and technology. Even with the foundation that these programs and activities supporting the day-to-day operating practices provide, the safe transportation of hazardous materials is complex due to a number of issues:

- Number of regulated hazardous materials (thousands are listed in regulations worldwide)
- Regulations that vary by mode, region, and country
- Different hazards classes including toxicity, flammability, corrosivity, and reactivity
- Various modes of transportation including road, rail, marine (including bulk vessels), pipeline, and air
- Multiple packaging types including bulk and non-bulk
- Use of more than one mode during a shipment (intermodal)

- Complexity of the supply chain including multiple parties and changes of custody during transit
- Overlapping and potentially unclear responsibility of various parties
- Transport routes where the risk profile can change depending on proximity to the public or other sensitive areas

The complexity of these activities, even with the current safety regulations and operating practices, is one of the reasons that accidents involving the transportation of hazardous materials occur. Table 1.1 illustrates some recent chemical transportation accidents in various countries and across different modes of transportation around the world.

Table 1.1 Examples of Recent Worldwide Chemical Transportation Accidents

Mode	Location and Date	Chemical Released	Description
Air	United States (7 August 2004)	Lithium-Ion Batteries	A fire destroyed some freight in a container that included lithium-ion batteries at the FedEx hub in Memphis, Tennessee. The freight container had been raised on loading equipment and pushed about halfway onto an airplane bound for Paris, France, when the loading personnel smelled smoke. They stopped loading and removed the smoking freight container.
Marine	Spain (13 November 2002)	Heavy Fuel Oil	The tanker Prestige, laden with 77,000 tons of heavy fuel oil, broke in two off the coast of Galicia (Spain), spilling an estimated 63,000 tons. Approximately 1,900 km of shoreline was affected in Spain and France. Around 141,000 tons of oily waste was collected in Spain and some 18,300 tons in France.
Pipeline	Belgium (30 July 2005)	Natural Gas	A pipeline carrying natural gas from the Belgian port of Zeebrugge to northern France exploded in Ghislenghien, Belgium, resulting in 24 fatalities and over 120 injuries. The cause of the accident was damage to the underground high-pressure pipeline from construction activities.
Rail	United States (6 January 2005)	Chlorine	A freight train traveling through Graniteville, South Carolina, encountered an improperly laid switch that diverted the train onto an industry track where it struck a parked train. The collision derailed 16 cars of the moving train. One car was breached, releasing chlorine gas. Nine people died as a result of chlorine inhalation, with an estimated 250 injuries and evacuation of about 5,400.

Mode	Location and Date	Chemical Released	Description
Road	China (26 October 2008)	Toxic Oil (Creosote)	More than 28,000 people in northern China had their water supply cut off after a truck carrying wash oil experienced brake failure, and overturned by the Yangjiapo reservoir in Shanxi province. About 30 tons of toxic oil were spilled, contaminating two million cubic meters of the reservoir.

In addition to safety considerations, security has become an increasingly greater concern for hazardous materials during transport. Certain hazardous materials could potentially be used as a weapon to spread fear, cause injuries or fatalities, and/or result in negative economic impacts. To address the potential for deliberate misuse, additional security regulations and industry standards continue to be developed.

To ensure the safety and security of chemicals in transit, this guideline promotes a comprehensive risk management program covering the entire supply chain. This holistic approach to risk management encompasses involvement from all stakeholders, and includes a general risk management framework highlighted in Figure 1.2.

The framework is designed to be flexible, enabling companies to tailor their TRM process to their individual circumstances and specific applications. The TRM framework is comprised of the following steps:

- *Primary Management System* is the foundation. Ensuring these elements are complete and current is the first step in the process. With the potential for cross-border shipments and international operations, understanding and keeping current with changes in regulations is critical. Chapter 2 presents the primary management elements for transportation of hazardous materials, including regulatory compliance, standards and guidelines, operational management, emergency preparedness and response, incident reporting, management of change, and auditing.
- *Identification and Prioritization* consists of (1) cataloging the hazardous materials and modes of transportation, (2) identifying sensitive areas and potential points of failure along the transit route, and (3) understanding interactions with other stakeholders in the supply chain. This step is a key in the TRM process for the identification of shipments that may require special attention, including additional resources for further evaluation and risk management. Prioritization, along with other risk assessment fundamental concepts, is presented in Chapter 3.
- *Risk Analysis* has been divided across several chapters in this guideline, each focusing on different levels of safety and security analysis. This separation is intended to help users efficiently locate information tailored to the needs and complexity of specific issues. A complete overview of the main risk analysis concepts is covered in Chapter 3. These fundamental concepts are applicable to people new to TRM, or those needing a refresher. Subsequent chapters present, many techniques, from simple, quick, and general to detailed, time-

consuming, and quantitative assessments. Simpler techniques should be the starting point with the more complex and resource-intensive activities conducted on a risk-justified basis. Chapter 4 presents qualitative and semi-quantitative approaches applicable to issues escalated through prioritization. For those issues that require additional detail for decision-making, Chapter 5 summarizes quantitative risk analysis techniques, with the original Guidelines book (CCPS, 1995) continuing to serve as the main reference for quantitative transportation risk analysis. Chapters 4 and 5 are specific to safety concerns, with the unique issues of security risk analysis presented in Chapter 6.

Figure 1.2 TRM Framework

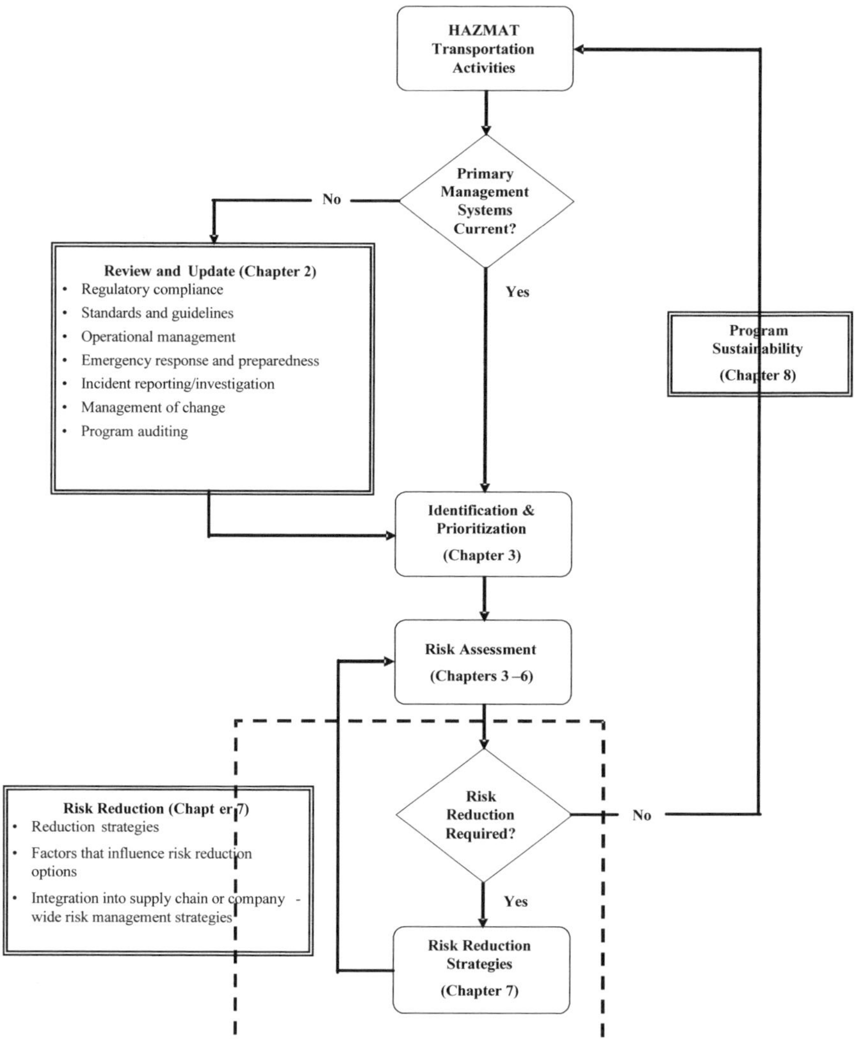

- *Risk Reduction* is the process of determining if the assessed risk is being managed adequately and evaluating/implementing options to reduce risk, as needed. If the risks merit further reduction, methods can be developed for the hazardous material operation, coordinated with the enterprise risk management philosophy of a single company, and/or developed in coordination with other supply chain partners. Chapter 7 discusses the strategy for identification, evaluation, and selection of risk reduction options, as warranted.
- *Program Sustainability* is the final step in the TRM framework and is designed to ensure the risk management process remains current. Presented in Chapter 8, this step includes an ongoing commitment to managing risk, continuously improving, reacting to emerging trends, and keeping current on evolving transportation risk analysis practices.

1.3 USING THESE GUIDELINES

While *Guidelines for Chemical Transportation Safety, Security, and Risk Management* has been developed for a broad range of transportation practitioners, the use of the concepts in this book will differ depending on one's role in the transportation supply chain or the transportation issue being assessed. Table 1.2 is a guide for using this book and presents chapters that are anticipated to be the most relevant for specific roles and responsibilities. Other chapters should be of interest depending on the level of detail required in the risk analysis, on other needs, or on the specific interests of the reader.

Table 1.2 Relevant Chapters of the Guidelines

	Transportation Professional								
Chapter	Logistics Service Providers	Shippers & Customers	Business Unit Managers	Transportation Managers	Safety	Security	Risk Managers	Government Regulators	Industry Associations
1–Introduction	x	x	x	x	x	x	x	x	x
2–Primary Management Systems	x	x	x	x	x	x	x	x	x
3–Risk Assessment Fundamentals	x	x	x	x	x	x	x	x	X
4–Qualitative and Semi-Quantitative Risk Analysis	x	x	x	x	x	x	x		
5–Quantitative Risk Analysis				x	x		x		
6 –Security Considerations	x	x	x	x	x	x	x	x	X

	Transportation Professional								
Chapter	Logistics Service Providers	Shippers & Customers	Business Unit Managers	Transportation Managers	Safety	Security	Risk Managers	Government Regulators	Industry Associations
7 –Risk Reduction Strategies	x	x	x	x	x	x	x	x	X
8–Program Sustainability	x	x	x	x	x	x	x	x	X
Appendices	x	x	x	x	x	x	x	x	x

REFERENCES

Canadian Broadcasting Centre. *Oil spill cuts water to nearly 30,000 people in China*. CBC News Page dated November 1, 2006. www.cbc.ca.

Center for Chemical Process Safety, *Guidelines for Chemical Transportation Risk Analysis,* New York: American Institute of Chemical Engineers, 1995

National Transportation Safety Board, *Cargo Fire Involving Lithium-Ion Batteries, Memphis, Tennessee August 7, 2007,* Hazardous Materials Accident Brief, NTSB/HZB-05/0, 2005

National Transportation Safety Board, *Collision of Norfolk Southern Freight Train 192 with Standing Norfolk Southern Local Train P22 With Subsequent Hazardous Materials Release at Graniteville, South Carolina, January 6, 2005,* Railroad Accident Report, NTSB/RAR-05/04, 2005

VROM Inspectorate, *2005 Annual Report*, The Netherlands Ministry of Housing, Spatial Planning and the Environment, 2006

Wadsworth, T., *Evaluation of the Response by Specialized Foreign Vessels to the Release of Oil From Prestige.* ITOPF Ltd., 2005

2
PRIMARY MANAGEMENT SYSTEMS

This chapter discusses the baseline programs for safety and security management for all modes of hazardous material transport. These baseline or "primary" management systems are those things that need to be in place prior to an organization being ready to attempt a risk analysis. Subsequent chapters of this guideline will focus on the next tier of risk management activities. The first tier of risk management activities for all companies is to ensure that their baseline programs are in place, complete, and current.

All members of the supply chain should have primary management systems in place to address compliance with all applicable regulations and implementation of additional programs and practices to manage the safety and security of hazardous material transportation activities. An overall transportation safety and security management system is recommended to assist in:

- Ensuring regulatory compliance
- Establishing and meeting training requirements
- Establishing procedure and protocols
- Ensuring equipment integrity (containers, rolling stock, pipelines, etc.)
- Monitoring and improving practices for the supply chain
- Managing changes to transportation procedures and practices
- Preparing for incidents and emergencies
- Identifying and managing transportation risks

It is difficult to develop a comprehensive list of primary systems that will meet every need and suit all situations. However, it is important to recognize that the success of any initiative or program is dependent on an existing management system, the corporate culture, and the continued support from management. Figure 2.1 illustrates that one's daily practices should be built on company policies and standards, while ensuring that all applicable codes and regulations are followed.

Figure 2.1 Primary Management Structure

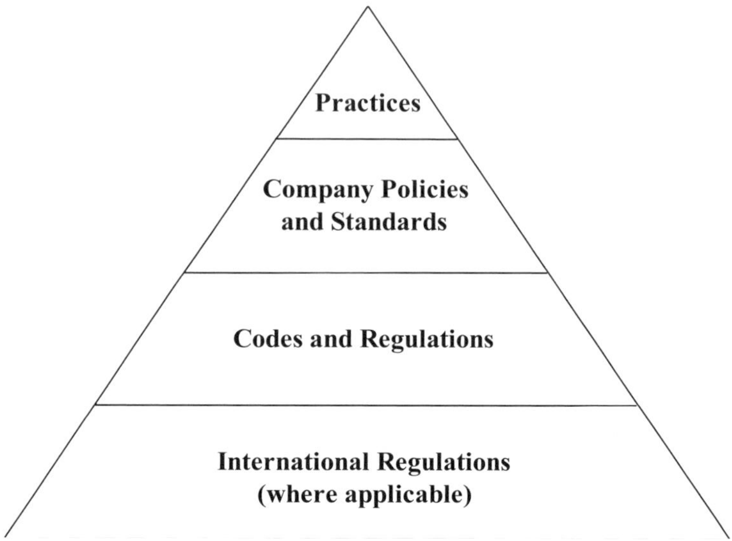

Where practicable, transportation risk management activities should be integrated into existing management systems including those for safety, process safety management, security, environmental stewardship, quality assurance, and business proficiency. Existing systems may provide a firm foundation for incorporating transportation risk management program elements and regulatory requirements. Many times, however, in the area of transportation activities, regulatory, safety, and security improvement programs are handled in different parts of the organization. Therefore, if integration is not feasible or practical, at least some level of connectivity should be established including an active and sustainable communication process.

2.1 REGULATORY COMPLIANCE

The transportation and distribution of chemicals are governed by a wide range of international and regional regulations. Since these regulations continue to evolve, it is important for companies to stay current with safety and security requirements in the countries where they and their supply chain partners operate. In the end, a clear and current understanding of the applicable regulations and interactions will be required to ensure operating compliance.

2.1.1 UN Model Framework

A widely applied regulatory framework concerning the transport of hazardous materials (or dangerous goods as they are more commonly referred to internationally) is presented in the "Model Regulations on the Transport of Dangerous Goods." This model has been developed by the United Nations Economic and Social Council's Committee of Experts on the Transport of Dangerous Goods in light of technical progress, the advent of new substances and materials, the necessity of modern transport systems, and, above all, the requirement to ensure the safety of people, property, and the environment. Although only a recommendation, the UN Model Regulations have been drafted in the mandatory sense (i.e., the word "shall" is employed throughout the text rather than "should") in order to facilitate direct use as a basis for national and international transport regulations. The model regulations aim at presenting a basic scheme of provisions that allow uniform development of national and international regulations governing the various modes of transport, yet attempt to remain flexible enough to accommodate special situations.

They are addressed to governments and international organizations concerned with the regulation of the transport of dangerous goods and include the following elements:

- *General Provisions, Definitions, Training, and Security* includes both safety and security training, and provisions for high-consequence dangerous goods, including the development of security plans
- *Classification* defines the hazardous materials on the basis of their properties (e.g., toxic, flammable, explosive, corrosive)
- *Dangerous Goods List* contains relevant information on the listed hazardous materials including the UN number, chemical name and description, hazard class, packing group(s), label(s), special provisions, exemptions, and limited quantities
- *Packing Provisions* include instructions and special provisions for each chemical on the dangerous goods list
- *Consignment Procedures* consist of proper marking, labeling, placarding, and documentation
- *Conditions of Carriage, Loading, Unloading, and Handling* includes additional requirements based upon the classification of the dangerous good

The model regulations also include requirements for the construction, testing, and approval of vehicles and packaging. If your company or supply chain partners perform these activities, then the appropriate section of the UN Model Regulations should be reviewed.

2.1.2 International Regulations

There are three international regulations that cover hazardous material transportation operations around the world, and specifically apply to the maritime and aviation modes of transport. The following international organizations implemented the international regulations for maritime and aviation:

- International Maritime Organization
- International Civil Aviation Organization
- International Air Transport Association

Since the ownership, management, and operation of ships and aircraft can span many different countries, often transiting outside their country of registry and jurisdiction, international standards that can be adopted, accepted, and followed by all shipping nations are critical.

2.1.2.1 International Maritime Organization

Maritime transportation is covered by the International Maritime Organization (IMO). Specific to hazardous material transportation, IMO has developed the International Maritime Dangerous Goods (IMDG) Code for transportation on the high seas in conjunction with government regulations under the International Convention for the Safety of Life at Sea (SOLAS). The IMDG Code contains internationally agreed-upon guidance on the safe transport of dangerous goods by sea and most commonly relates to the carriage of dangerous goods in freight containers and tank containers, with particular reference to the segregation of incompatible substances.

Maritime security is also an integral part of IMO's responsibilities. A comprehensive security regime for international shipping went into effect on July 1, 2004. The mandatory security measures, initially adopted in December 2002, include a number of amendments to the 1974 SOLAS, the most far-reaching of which enshrines the new International Ship and Port Facility Security (ISPS) Code, which contains detailed security-related requirements for governments, port authorities, and shipping companies.

2.1.2.2 International Aviation Regulations

The International Civil Aviation Organization (ICAO) has developed regulations for air transport of hazardous materials that are based upon the UN Model, but modified to accommodate the unique aspects of air transport. Annex 18, dealing with the Safe Transport of Dangerous Goods by Air, in general sets down broad principles, but one of the standards requires that dangerous goods are carried in accordance with the *Technical Instructions for the Safe Transport of Dangerous Goods by Air*, known as the Technical Instructions. The Technical Instructions contain a very comprehensive set of requirements; among other things they provide for the classification of dangerous goods and list those goods that are:

- Forbidden under any circumstances
- Forbidden on both passenger and cargo aircraft in normal circumstances but could be carried in exceptional circumstances subject to exemption by the states concerned
- Forbidden on passenger aircraft but permitted on cargo aircraft in normal circumstances
- Permitted on both passenger and cargo aircraft in normal circumstances

Working closely with governments in the development of the regulations, including ICAO and other national authorities, the International Air Transport Association (IATA) ensures that the rules and regulations governing dangerous goods transport are both effective and efficient. Through its dangerous goods regulations (DGR) and a comprehensive and effective training program, IATA makes certain that shippers, forwarders and carriers have the tools and resources to ship dangerous goods safely.

Based on international and national air regulations, as well as airline-specific requirements, the DGR guide helps shippers of dangerous goods with the following requirements:

- Classification
- Marking
- Packing
- Labeling
- Documenting

2.1.3 Regional Regulations

Depending on the country of operations and the mode of transport, many different regulations may apply for the safe and secure transportation of hazardous materials. This section details the regional hazardous material regulations in Europe and North America. Appendix C, on the enclosed CD, can be used as a resource for specific regulations in other countries and emerging markets. Since many of the evolving safety and security regulations are based on the UN Model Regulations and those of Europe and the United States, these regulations provide a good background of the different requirements.

However, because transportation safety and security regulations continue to develop and evolve, it is critical to review and understand all current regulatory requirements in the countries of operation.

2.1.3.1 Europe

In Europe, there are specific regulations that are followed for road, rail, and inland waterway transport of hazardous materials. The international regulatory schemes are based upon the UN Model Regulations and include:

- *ADR for Road Transport*—The European Agreement concerning the International Carriage of Dangerous Goods by Road (ADR stands for *Accord européen sur le transport des marchandises dangereuses par route*) is intended primarily to increase the safety of international transport by road, but it is also an important trade facilitation instrument. While participation in the ADR agreement is not limited solely to European Union countries, the requirements of ADR have been annexed to the European Union Council Directive 94/55/EC on the approximation of the laws of the Member States with regard to the transport of dangerous goods. Therefore, these requirements have become applicable not only to international transport of dangerous goods, but also to national traffic in all countries of the European Union. The Member States implement these EU directives into their own regulatory framework, where the country-specific regulations take precedence. This is particularly important when country-specific regulations contain additional or more restrictive requirements.

- *RID for Rail Transport*—RID is the abbreviation of "Reglement International concernant le transport des marchandises Dangereuses par chemin de fer," which translates as Transport of Dangerous Goods by Rail. These regulations define the required type of construction of the tank cars and the markings for transport.
- *ADN for Inland Waterway Transport*—The ADN (European Agreement Concerning the International Carriage of Dangerous Goods by Inland Waterways) aims to ensure a high level of safety, contribute to the protection of the environment, facilitate transportation operations, and promote the international trade of chemicals.

2.1.3.2 North America

This section provides a summary of the regulations for the United States, Canada, and Mexico.

In the United States, the Department of Transportation (U.S. DOT) is the primary regulatory body for transportation-related regulations. The Hazardous Materials Transportation Act (HMTA) authorizes U.S. DOT to regulate the transportation of hazardous materials. The Pipeline and Hazardous Materials Safety Administration (PHMSA) has responsibility for safe and secure movement of hazardous materials by all transportation modes, with the exception of bulk marine shipments. PHMSA establishes the following hazardous materials regulations, which are contained in Title 49 of the Code of Federal Regulations (CFR):

- *Hazmat Safety Regulations*—Title 49 CFR Parts 100–185. Specific sections by mode of transportation include:
  - Part 174: Carriage by Rail
  - Part 175: Carriage by Aircraft
  - Part 176: Carriage by Vessels (Marine)
  - Part 177: Carriage by Public Highway (Road)
- *Pipeline Safety Regulations*—Title 49 CFR Parts 190–199

In addition to PHMSA, U.S. DOT has a number of mode-specific agencies that work with PHMSA in the development of hazardous material regulations, although PHMSA has the primary responsibility for the safe and secure movement of hazardous materials by all transportation modes, including the nation's pipelines. Other agencies that are responsible for implementing and enforcing the regulations include:

- *Road Transport*—Federal Motor Carrier Safety Administration (FMCSA)
- *Rail Transport*—Federal Railway Administration (FRA)
- *Air Transport*—Federal Aviation Administration (FAA)

In the security arena, the Transportation Security Administration (TSA) of the Department of Homeland Security, working in a joint, collaborative effort with the U.S. DOT, focuses on establishing, implementing, and enforcing hazardous materials transportation security requirements.

The United States Coast Guard (USCG) has the responsibility of maritime safety and security. The USCG Office of Operating and Environmental Standards develops and maintains standards regulating marine industry through treaties, regulations, and policy, and represents the United States in national and international organizations, including the International Maritime Organization (IMO). The USCG is responsible for the enforcement of the Maritime Transportation Security Act of 2002 (MTSA), signed on November 25, 2002. This law is the U.S. equivalent of the international ISPS Code, and was fully implemented on July 1, 2004. In addition to regulations, the USCG uses Navigation and Vessel Inspection Circulars (NVICs) to provide detailed guidance regarding marine safety and security regulations and programs.

In Canada, the Transport Dangerous Goods (TDG) Directorate promotes public safety during the transportation of dangerous goods. The TDG Directorate serves as the major source of regulatory development, information, and guidance on dangerous goods transport for the public, under the Transportation of Dangerous Goods Act (TDG Act), implemented in 1992. While the TDG Act defines the coverage of the requirements, the prohibitions, and the authority to make regulations, the TDG regulation includes the technical requirements and references to standards. The regulation is comprised of 16 parts, with specifics for different modes of transportation found in:

- Part 9: Road
- Part 10: Rail
- Part 11: Marine
- Part 12: Air

The Official Mexican Standards (Normas Oficiales Mexicanas or NOMs) augment the Mexican Regulation for the Land Transport of Hazardous Materials and Wastes. The Mexican Secretariat for Communications and Transport (SCT) is responsible for publishing and maintaining the NOMs. In addition, other Mexican government agencies have published standards relevant to the transportation of hazardous materials within Mexico. The Mexican hazmat NOMs are fairly consistent with those of the UN Model Regulations.

2.1.4 Harmonization of Transportation Regulations

Most national and international regulatory frameworks for hazardous material transportation are harmonized to a greater or lesser degree with the UN Model Regulations. Governments, intergovernmental organizations, and other international organizations, when revising or developing regulations for which they are responsible, are expected to conform to the principles set forth in the UN Model Regulations, thus contributing to worldwide harmonization in this field.

The United States, along with several other developed nations, participates in a number of international forums as part of the ongoing process of harmonizing with international standards and regulations. Participation ensures that the interests of each nation are communicated and considered in the development of international standards, and that the continued evolution of the national regulations will move toward conforming directly to the UN Model. An example is the Globally Harmonized System for the Classification and Labeling of Chemicals (GHS). By promoting common, consistent criteria for classifying chemicals and by developing compatible labeling and safety data sheets, the Globally Harmonized System is intended to enhance public health and environmental protection, as well as reduce barriers to trade.

2.2 ESSENTIAL COMPONENTS OF A TRANSPORTATION MANAGEMENT SYSTEM

Full compliance with the regulatory system is clearly a minimum requirement that all supply chain partners must always strive to achieve and maintain. Many companies have implemented additional internal procedures and processes to provide a more holistic approach to transportation risk management than is required by regulation. Because the regulatory system serves as the foundation in an overall transportation risk management program, this foundation must be in good shape before working on enhancing other elements of your company's transportation risk management program.

A transportation risk management system should be based on the following principles:

- *Commitment*—There must be a tangible, visible commitment—including resources—from management and the work force to reduce risks. Risk management should be everyone's job. Provide incentives to reinforce the commitment. Be accountable.
- *Culture*—Promote a proactive "risk reduction culture" in day-to-day operations. Ask risk questions when making decisions and performing operations. Incorporate risk considerations into basic management systems, such as recordkeeping, quality control, performance evaluation, and training. Think risk reduction.
- *Collaboration*—The most effective risk management is built on interaction among all parties involved in a hazardous material transport chain (e.g., shipper, package manufacturer, carrier, and consignee). Don't try to manage risks in a vacuum. Team up to manage risk effectively.
- *Prioritization*—Because there typically are numerous risks to address and various ways to reduce them, and both public and private resources for managing risks are limited, priorities must be set. Establish priorities, based on analysis, to address the worst risks first. Revisit your priorities as your business changes. Articulate your risk reduction priorities.
- *Action*—Risk is reduced by concrete actions specific to your hazardous materials transport operations. Actions are based on risks, costs, and

benefits, factoring in such realistic considerations as technical feasibility, ease of implementation, budgets, competition, regulatory burden, and legal constraints. Action is the heart of effective risk management. Planning and analysis, while necessary, do not reduce risk. Actions do. Adopt a bias for action.

- *Continuous Improvement*—Not all risks associated with hazardous materials transportation can be totally eliminated. Through commitment, self-evaluation, and the flexibility to change, improvements in risk management results and efficiency should be sought continuously. Adapt to get better.
- *Communication*—All parties who have a role in risk management—including company management, employees, consignees, suppliers, and emergency responders—need to know their role and be aware of relevant risk information (e.g., nature and level of risk, risk control points). Appropriate documentation and dissemination of risk analyses and risk reduction strategies can facilitate communications. Share risk knowledge.

A critical element of transportation risk management is management leadership and commitment. There must be a tangible, visible perceived commitment—including resources—from management to reduce and manage transportation-related risks. This commitment can be established via written policies stating that commitment, the establishment of accountability by personnel responsible for risk management program activities, the formation of an organizational structure to support the risk management program, and the establishment of risk management partnerships with other parties in the supply chain (e.g., package manufacturers, carriers, toll processors, warehouses, distributors). Management needs to dedicate resources to conduct risk assessments and eliminate unacceptable risks. Visible management commitment will promote a proactive "risk reduction culture" where questions about risk are considered in making day-to-day decisions and performing routine operations.

In addition to management commitment and a culture of risk management, the following activities are part of a comprehensive risk management system:

- Industry standards and guidelines
- Operational management
- Hazard and risk assessment
- Emergency preparedness and response
- Incident reporting and investigation
- Management of change
- Auditing

Each of these elements is discussed in detail in the following sections.

2.2.1 Industry Standards and Guidelines

The two primary industry consensus programs for managing transportation risk are the Responsible Care® Distribution Code of Management Practices and the National Association of Chemical Distributors' Responsible Distribution Process^{SM} (RDP). These programs share similar elements and are widely used across the chemical distribution supply chain. Responsible Care® is a recognized internationally performance-based system that contains the following elements:

- Distribution risk management
- Compliance review and training
- Carrier safety
- Handling and storage
- Emergency preparedness

Transportation security is also an important element of a transportation risk management program, and a number of government regulations and industry consensus guidelines have now been established. The Responsible Care® Security Code provides a set of management practices for ensuring the security of hazardous material transportation operations and also calls for application of these practices throughout the supply chain.

Application of Responsible Care® efforts for the safe and secure transport across the supply chain should not take a one-size-fits-all approach. The efforts should be tailored to the hazards of the materials, modes of transportation, complexity of the supply chain, and the identified and potential risks of the operation.

2.2.2 Operational Management

Many of the elements of a transportation risk management system must be built into the ongoing management of transportation operations. These activities start with the chemical shipper, but must also extend to the entire supply chain network. Each shipper's location should have procedures and practices in place to properly load, secure, and present the shipment for transportation. These may include:

- Having a securement policy that includes and safety and security pre-loading inspections and post-loading inspection checklist
- Reviewing shipping papers for completeness
- Ensuring proper placarding
- Ensuring the Emergency Response Plan (both on site and en route) is current
- Ensuring emergency response capabilities, as may be required for the material and mode of transportation
- Implementing and conducting appropriate training programs
- Observing activities of third parties to ensure that safety and security practices are followed
- Having a documented process for providing feedback third parties

As part of operational management, a procedure should be developed for qualifying carriers. Carrier selection should include regulatory compliance, safety programs, security, and performance, with qualification based on the following steps:

- Determine the potential carriers that are qualified for the operation.
- Collect and evaluate regulatory compliance data for potential, qualified carriers such as notices of violation (NOV), fines, and penalties.
- Collect and evaluate performance data on the health, safety, security, environment, and transportation programs of each potential, qualified carrier.
- Select a list of approved carriers based on the performance data and on the quality of the carrier management system that is delivering that performance.
- Evaluate transportation routes, delivery timing, and equipment selection (container size and type).

A similar qualification, selection, and review process should be instituted for other components of the supply chain, such as distributors, warehouses, terminals, toll processors, transloading locations, and container cleaning facilities. Inherent in this process is the need to ensure these third-party handlers are provided the necessary information about the chemicals they transport, handle, and/or store, and requirements to manage their operations safely and securely.

2.2.3 Emergency Preparedness and Response

Transportation of hazardous materials involves inherent risks, not all of which can be fully controlled. Incidents have occurred in all segments of the supply chain; therefore, a program for preparing for and responding to emergencies involving hazardous materials in transport needs to be in place before a transport shipment commences. The program must comply with all applicable regulations and may include elements such as procedures, response equipment, training and maintenance on equipment, response capability needs and special concerns, emergency response plans, and coordination and communication with emergency agencies and the public.

In North America, hazardous material shippers are required by regulation to provide an emergency contact phone number on the shipping papers for emergency response agencies to obtain technical advice in the event of an accident or emergency en route. Many companies elect to register with a third-party supplier, such as CHEMTREC in the United States, or CANUTEC in Canada, to comply with this requirement. The company chosen must maintain a database of each shipper's hazardous materials information, as well as contact information for designated industry personnel who can provide special expertise on individual materials.

Many third-party suppliers also contract with companies to handle on-scene emergency response if the company does not have, or chooses to supplement, their own internal resources. When operating outside the United States and Canada, a local emergency response provider should be identified for specific countries of operation.

2.2.4 Incident Reporting and Investigation

The adage, "what gets measured gets improved" applies to transportation risk management. Companies should establish internal reporting systems for all transportation-related accidents/incidents.

The information derived from reports of transportation incidents provides an excellent outcome measurement of the performance of a company's transportation safety management program. Information about the causes of transportation incidents not only provides the basis for proper corrective actions, but also provides useful lessons for the future, which can lead to improvements in the transportation safety management system.

Incident reporting should be mandated across the supply chain, and the need and benefit of such reporting should be communicated across the supply chain. Periodic meetings should be held with the supply chain partners to review incidents and lessons learned.

All transportation-related incidents, even ones that do not result in any damage, injuries, security breaches, or environmental impacts (including near misses), should be reported and investigated and corrective actions taken to prevent recurrence. Since it may not be feasible, however, to investigate every incident to the same degree, a definition of major incidents and high-potential incidents that will be investigated in detail (incidents which under different circumstances could have been major incidents) must be developed. Possible criteria include accidents that arouse public attention including media coverage (e.g., large spills, fires/explosions, fatalities, and inconvenience/evacuation) and accidents that may involve the shipping and/or transportation company in litigation. Because major incidents can have wide-ranging effects on a shipping company's, as well as downstream supply chain partners' reputation, license to operate, and potential litigation, it is extremely important that all possible actions be taken to mitigate the effects of major incidents and to prevent their recurrence.

As with investigation programs for other types of incidents, the full value of a management system for transportation incident investigation procedures can only be realized if there is proper follow-up of the investigation, resolution of problems that come to light, and action on recommendations. As part of these programs, the shipping company should confirm that the transportation incident was properly investigated and that follow-up recommendations have been resolved by the responsible party. In some cases, the shipper/offerer may want to participate in the root cause investigation and visually inspect carrier records to confirm that corrective action was implemented (e.g., a photo or provision of a new procedure/training record). The management system for transportation incident investigation must have as one of its main objectives the timely and complete communication of results of incident investigations and corrective actions to all those who need to know, including other carriers and other members of the supply chain. Sharing lessons learned and corrective actions will help ensure that similar incidents can be avoided in the future. In all cases, an incident report should be generated which includes, at a minimum:

- Date
- Description of the incident
- Factors that contributed to the incident
- Recommendations to prevent future occurrences

The information gathered on transportation incidents has long-term value in addition to the immediate lessons learned. For example, this information may be useful for monitoring performance over time. It can be used to build up a database on transportation incidents. Finally, and most importantly, such information can provide statistical data for transportation risk assessments and improvement efforts. Trend analysis of causes of transportation incidents is vital to a management system that addresses major problems and sets priorities for risk reduction. This element should be part of each supply chain partner's transportation risk management. These risk reduction strategies are discussed in detail in Chapter 7.

Although investigation of transportation incidents is, in some ways, different than investigation of process incidents, many of the same principles apply. A full description of techniques and programs for incident investigation can be found in *Guidelines for Investigating Chemical Process Incidents* (CCPS, 2003).

2.2.5 Management of Change

Management of change applies to transportation risk management in a similar manner to process risk management. Transportation-related changes should be managed, including those associated with any of the following categories:

- Changes in equipment or packaging
- Changes in business (e.g., a new supply chain partner, a new material to be transported, a new loading location, or a new delivery point)
- Changes in routing
- Changes in transportation mode
- Changes in procedures (including loading/unloading, maintenance/inspection, and security procedures)
- Changes in regulations or changes in activities, products, and services that cause the list of applicable regulations to change

A system for managing changes (including temporary changes) should be instituted as part of the transportation risk management system. The intent of a change management system is to ensure that any new or modified risks associated with a modification are identified, understood, and minimized, managed, or eliminated. In all cases, the risk implications of changes should be examined and understood before the change is implemented.

A change management system should include procedures for proper documentation of the change and for communication of the change to and training of necessary personnel. Documentation and training helps to ensure that all affected employees are made aware of the changes and the steps or precautions taken to minimize any resulting risk. A change management system should include the following elements:

- Definition of what constitutes a change (including what constitutes an "in-kind replacement")
- Documentation of the proposed change
- Review of potential safety, security, health, environmental, and compliance-related impacts
- Physical inspection to ensure readiness of equipment and personnel prior to implementation (pre-startup review)
- Resolution of safety, security, health, environmental, and compliance-related concerns generated during all stages of review
- Tiered approval system in which the level of management approval required is dependent upon level of risk
- Effective date for permanent change
- End date if temporary

The change management system should include a change management form for use in documenting all relevant information.

2.2.6 Transportation Risk Management System Auditing

A good transportation risk management system includes periodic audits of the management system, its technical elements, and compliance across the supply chain. The purpose of an audit program is to provide feedback on transportation risk management efforts to improve or enhance current operations such as determining:

- Compliance performance with:
  - Applicable hazardous materials regulations across the supply chain
  - Transportation risk management program procedures of the supply chain partner commissioning the audit and those of the supply chain partner being audited
  - Contract requirements
- Verification of supply chain partner qualifications
- Complete and up-to-date policies, procedures, and training
- Status and effectiveness of transportation risk management efforts versus goals or progress toward goals

The results of an audit should be reported to management, not only for their information but also to act as a device for accountability and action. As part of an audit program, how the output of the audit activity is to be documented and distributed must be determined. Perhaps all audits over a certain time period are summarized for management, or perhaps the results of any single audit may be reported to the person responsible for that activity. The goal should be to bring to the attention of the commissioning management and to the management of the supply chain partner being audited the issues illuminated by the audit process. Further, the audit program should extend the appropriate parties an opportunity to review and respond to these issues and, most importantly, to formulate action plans to remedy those issues requiring attention.

Reducing transportation risk through properly implemented corrective actions is a key benefit sought in audit programs. Corrective actions either can be part of the formal audit report or may be handled separately, depending on the management system being employed. In either case, assignment of responsibility, action plans for resolving the recommendations, and a follow-up system to verify completion are essential elements in capturing the value opportunities created through audit programs.

2.3 XYZ CHEMICAL EXAMPLE—PRIMARY MANAGEMENT SYSTEMS

This chapter introduces XYZ Chemical, the company that will be used to illustrate the concepts presented in this book. XYZ Chemical primarily engages in pesticide and herbicide manufacturing operations in the United States, Europe, Asia, and South America. A corporate-level review of the chemical hazards is presented in Chapter 3, and the review focuses on the hazardous material transportation operations of a single Asian facility in Chapters 4–8. The specific chemicals and operations are detailed in the later chapters.

XYZ Chemical has brought in a new manager of risk management who has begun an initiative to develop a formalized global risk management program. While the company has a strong safety and security management program in Europe and North America, the focus of the example is on ensuring that the primary management systems and programs in the company's other operating regions meet the corporate environmental, health, and safety (EHS) standards, as well as integrating all operations into a centralized risk management process.

Once the overarching risk management program is finalized and implemented across the company, the primary management system review will be conducted at the facility level. Since this is the implementation stage of the program within XYZ Chemical, this activity is conducted as a corporate-level review. From this review it was determined that a formal risk management system has been developed and implemented in the United States, but only ad hoc risk evaluations are used in the other operating regions.

Based on this finding, the following recommendations were implemented before any formal risk analysis of XYZ Chemical as a whole, any individual facilities, or any of its transportation operations was conducted:

- The formal risk management system in the United States will be upgraded and rolled out as the corporate standard for all operating regions, including all of North America, Europe, South America, and Asia.
- Once the corporate management system is complete, the baseline activities of all hazardous material transportation operations will be compared across all regions to ensure the same baseline level of safety and security, including:
  - Complying with regulations
  - Establishing and meeting training requirements
  - Establishing procedure and protocols
  - Ensuring equipment integrity (containers and rolling stock)
  - Monitoring and improving practices for the supply chain
  - Preparing for incidents and emergencies
  - Identifying and managing transportation risks

The resulting initial program for XYZ Chemical is a process for managing risk for their worldwide operations, including consistent primary management systems. Once this risk management system is in place, corporate management can begin a prioritization process (Chapter 3) to identify hazardous materials and specific facilities and transportation activities that require additional risk analysis.

REFERENCES

American Chemistry Council, *Guidelines for Managing Process Change*, 1993

American Chemistry Council, *Implementation Resource Guide for Responsible Care® Security Code of Management Practices: Value Chain Activities*, September 2002

Center for Chemical Process Safety, *Guidelines for Implementing Process Safety Management Systems*, American Institute of Chemical Engineers, New York, 1994

Center for Chemical Process Safety, *Guidelines for Investigating Chemical Process Incidents, Second Edition*, American Institute of Chemical Engineers, New York, 2003

Center for Chemical Process Safety, *Guidelines for Risk Based Process Safety*, American Institute of Chemical Engineers, New York, 2007.

Chemical Manufacturers Association, *A Guide to the Distribution Code of Management Practices*, Washington, D.C., 1991

Chemical Manufacturers Association, *Responsible Care®: Distribution Code of Management Practices*, Washington, D.C., 1990

United Nations Economic Commission for Europe, Recommendations on the Transport of Dangerous Goods Model Regulation Volume I, Fifteenth Revised Edition, United Nations, New York and Geneva, 2007

3
RISK ASSESSMENT FUNDAMENTALS

This chapter presents the basic concepts and definition of risk (Section 3.1), a protocol for conducting transportation risk assessments (Section 3.2), and a prioritization process for identifying important issues and transportation scenarios requiring a more detailed risk analysis (Section 3.3). Due to the differences in safety and security definitions and risk assessment methodologies, the focus of Chapters 3, 4, and 5 is limited to transportation safety. Security concepts, definition, and assessment methods are presented separately in Chapter 6, with this chapter providing a high-level comparison of safety and security.

3.1 SAFETY RISK ASSESSMENT CONCEPTS

Risk assessment is a process used to systematically identify, understand, and ultimately make informed decisions regarding the options available to manage risk and for continuous improvement of the transportation operation/network. The steps involved in a safety risk assessment, as illustrated in Figure 3.1, include:

- *Step 1—Identification and Prioritization*: screening utilized to identify and escalate issues/scenarios for more detailed risk analysis.
- *Step 2—Risk Analysis*: the process (qualitative or quantitative) of evaluating consequence and likelihood and estimating the overall level of risk associated with the selected scenarios.
- *Step 3—Risk Evaluation*: comparison of results against evaluation criteria used for making decisions regarding the need for risk mitigation.
- *Step 4—Risk Reduction*: development, comparison, and selection of risk mitigation options to reduce risk to a target level, if needed, or as needed.

Figure 3.1 Risk Assessment Steps

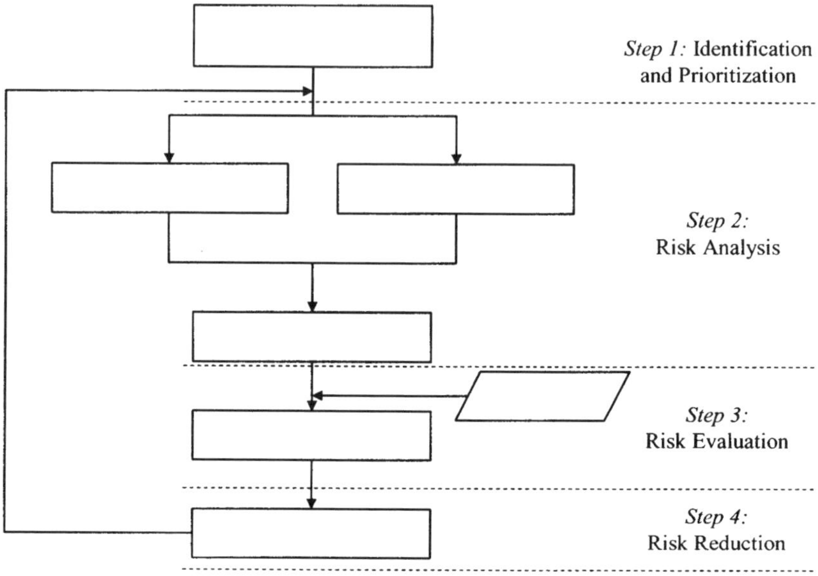

In addition to an overall description of risk assessment fundamentals, this chapter details the identification and prioritization process used to select scenarios for a more detailed risk analysis (Step 1). Risks identified as "lower priority" during the prioritization process should not be disregarded or ignored, but should be managed as part of an overall transportation safety and security management system. This management system for safety and security should include the primary management system program elements described in Chapter 2.

For issues escalated for further assessment (Step 2), a risk-based approach can help ensure that estimates of potential consequences, likelihood, and risk are evaluated consistently using standardized and accepted techniques. Using this approach, risks are evaluated using approved risk criteria and decisions are made using an objective approach (Step 3). While the goal is to make risk-informed decisions, to imply that companies will make decisions solely based on risk is unrealistic, as there may be other business, operational, or security trade-offs, or the company may decide to accept a slightly higher risk for the short term while long-term risk reduction strategies are put into place. This approach can be documented and applied consistently within or across an organization and includes the process for selecting risk mitigation options to achieve a desired level of risk reduction (Step 4).

Risk assessment of hazardous material transportation can ultimately lead to more informed decisions since the process enables a better understanding of the following items:

- Features of various modes of transport
- Types of incidents that might occur

- Characteristics of current and alternate routes
- Design and complexity of various distribution systems
- Depth and rigor of management systems (types/frequency of inspections/ maintenance, shipment tracking, stewardship efforts, etc.)
- Safety performance of carriers and other logistics service providers
- Possible options to reduce risks

3.2 RISK DEFINITIONS

Risk is a function of the consequence and likelihood of the specific transportation hazard scenario(s) under evaluation:

$$\text{Risk} = f(\text{scenario(s), consequence, likelihood})$$

In addition to the formal equation for risk, Table 3.1 defines specific terms as they are used throughout this book.

Table 3.1 Selected Transportation Safety Risk Definitions

Accident-Initiated Event: An event (or the first event in an event sequence) that is caused by a movement-related transportation accident (e.g., a train derailment or a barge grounding).
Consequence: The direct, undesirable result of an incident outcome, specifically the impacts resulting from a release of a hazardous material. Consequence is generally a function of the hazards of the material released, the extent of the release, and the presence of receptors (people, ecosystems, property, etc.).
Hazard: Inherent property or characteristic of a material, system, or process that has the potential for causing serious injury to people and/or property or environmental damage.
Incident: The loss of containment of material or energy (e.g., a puncture or fittings leak of ammonia on a railcar). Not all events propagate into incidents.
Incident Outcome: The physical manifestation of the incident: for toxic materials, the incident outcome is a toxic release, while for flammable materials, the incident outcome could be a boiling liquid expanding vapor explosion (BLEVE), flash fire, vapor cloud explosion (VCE), etc. (e.g., for a leak of chlorine from a railcar, the incident outcome is a toxic release).
Likelihood: A measure of the expected probability or frequency of occurrence of an event (e.g., events/year).
Non-Accident-Initiated Event: An event due to causes unassociated with the movement-related aspects of transportation (e.g., improperly closed valves or pressure build-up due to contamination).
Risk: A measure of human injury, environmental damage, or economic loss in terms of both the incident likelihood and the magnitude of the loss or injury.

While other terms and definitions may be used in specific transportation regulations and for other hazardous material activities, the definitions in Table 3.1 are consistent with those found in the *Guidelines for Chemical Transportation Risk Analysis* (CCPS, 1995) and *Guidelines for Chemical Process Quantitative Risk Analysis, Second Edition* (CCPS, 2000). These definitions are further defined and developed below.

3.2.1 Hazards

Identification of hazards and selection of initiating events are the first steps in any risk analysis. Hazards are generally related to the physical properties of the material being transported. The properties represent the inherent risk and the potential adverse consequences if the material were released. The hazards to be considered when evaluating different movements may include:

- Flammability
- Combustibility
- Toxicity
- Corrosivity
- Reactivity
- Others (e.g., explosivity, detonability, radioactivity, biological hazard, odor, asphyxiation, environmental persistence)

3.2.2 Initiating Events

In a safety risk analysis, initiating transportation events may include both accidents and non-accidents. Depending on the mode of transport, Table 3.2 details some of the types of accident-initiated events that should be considered in the risk analysis process.

Table 3.2 Accident-Initiated Events

Road	Rail	Waterway	Air	Pipeline
Collision Overturning Grade Crossing* Cargo Shifting	Collision Derailment Grade Crossing*	Collision Grounding Ramming Capsizing Allision	Crash Cargo Shifting	External Impact

**Grade crossing* means an accident at this type of location along a route

Accident-initiated events can generally be identified based on historical data for the industry or current operations/activities. Information from both actual events and near misses (high-potential events that did not result in a serious incident) should also be included to encourage an understanding of the full range of incident possibilities. Even though an event has not occurred does not mean that it is not credible, or should not be considered. Therefore, when brainstorming and developing accident scenarios, it is often helpful to work with carriers and other logistic service providers whose experience in their particular industry may help identify and validate the full range of potential events for further analysis.

The following are examples of potential causes, or contributors to, accident-initiated events, and may be considered across various modes:

- Equipment defects
- Human factors (incorrect or inappropriate response)
- Road/rail defects
- External events
- Navigational failures
- Control system failures
- Incorrect ballasting/balancing

There also needs to be consideration of how the packaging might perform if involved in an accident. As presented in Section 3.2.5, conditional release probabilities are applied in a risk analysis to represent the range of potential outcomes. These outcomes can range from no release to full package failure. While it is important to consider scenarios that can lead to a loss of containment, not all transportation accidents will result in a chemical release.

Non-accident-initiated events differ in that the event is not associated with an external impact during transport. Typical non-accident-initiated events that may be reviewed as part of a transportation risk analysis may not be mode-specific, and could include:

- Improper securement
- Corrosion
- Metallurgical failure
- Overpressure
- Equipment component failures (e.g., valves, rupture disks, fittings, distribution)
- Overfilling or underfilling
- Relief device activation due to surges
- Contamination
- Temperature changes (overheating, loss of refrigeration, freezing)
- Control system failures

3.2.3 Incident Consequences

The range of potential consequences of a chemical release is generally a function of the physical area of impact and the potential exposure to sensitive receptors (e.g., people, ecosystems, property) that may be located in the area of the release. The physical area involved is generally dependent on the following:

- Hazards and physical properties of the material
- Total volume shipped per container (bulk versus non-bulk)
- Shipping conditions (pressure and temperature)
- Container design and features
- Conditions of the accident
- Atmospheric conditions (e.g., ambient temperature, humidity, wind speed, and wind direction)

Depending on the hazards of the material, the release mechanism, and other factors, the release of a hazardous material from its transportation container may result in the following:

- Fire (chemical in transit or spilled fuel)
- Boiling liquid expanding vapor explosion (BLEVE)
- Vapor cloud explosion (VCE)
- Flammable vapor cloud
- Toxic cloud
- Asphyxiating cloud
- Environmental impact (e.g., chemical, biological, or radioactive contamination)

Figure 3.2 is a sampling of a range of outcomes that could occur given the release of a hazardous material. For example, if an accident involving a flammable liquid spill that ignites and partially or completely engulfs another transportation container, a BLEVE may occur, generating significant overpressure conditions (blast effects) and potential damage from equipment fragmentation. If the container contains a flammable material, the BLEVE may also result in a fireball with an associated thermal radiation hazard. If the release is a gas or high-pressure liquid and ignition occurs, a jet fire, fireball, and/or sustained fire may result. In the absence of immediate ignition, a large flammable vapor cloud or a liquid pool may form and, if ignited, a flash fire, vapor cloud explosion, or pool fire could result. If the material is toxic but nonflammable, a toxic hazard may result. A toxic hazard may also exist if the material has both toxic and flammable properties. Specifically when ignition does not occur, the toxic material is not completely consumed in the fire or explosion, or there are toxic by-products from a fire. Additionally, an environmental impact may occur depending on the location of the accident and proximity to sensitive receptors.

Figure 3.2 Potential Outcomes of a HAZMAT Transportation Event

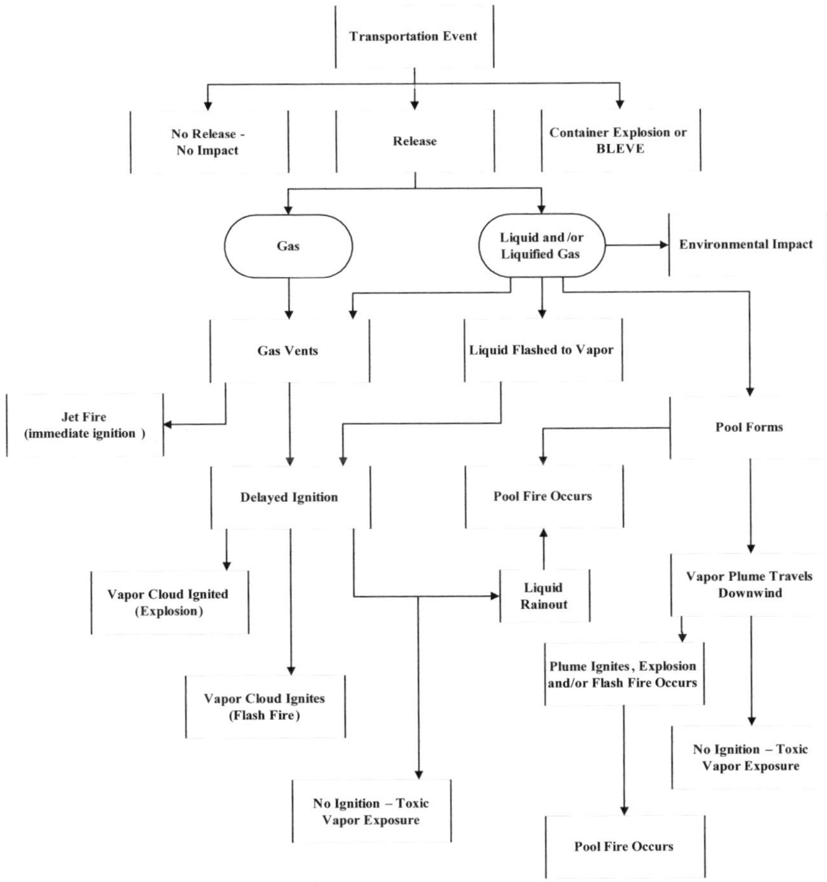

3.2.4 Incident Impacts

An incident can result in potential impacts to people, property, and the environment along the transportation route, including:

- Fatalities
- Injuries
- Property damage
- Environmental impacts
- Business interruption
- Evacuation
- Distribution system disruption (port or thoroughfare closures)
- Negative media attention
- Tighter governmental regulations

- Existing corporate reputation (i.e., previous incidents may impact "license to operate")

Factors that may be considered in evaluating the potential magnitude of the consequences include:

- Buffer zone along the route (e.g., transportation corridors and populated areas)
- Emergency preparedness and response capabilities, including community notification systems
- Prevailing meteorological conditions
- Terrain, including the presence of vegetation/trees, buildings, and other structures
- Population density and variation in population along the route
- Ability to shelter-in-place effectively (sealed buildings vs. open-air housing)
- Presence of sensitive receptors (e.g., schools, hospitals, large gathering places)
- Impact on existing corporate reputation

In addition to consequences that could affect people and property, a release could have an environmental impact. The potential for environmental impacts can be evaluated by mapping the transportation route, and considering transit through, over, or in proximity to:

- National or regional forests or parks
- Officially designated wildlife sanctuaries, preservers, or refuges
- National wilderness areas
- Major bodies of water, including recreational areas
- Drinking water aquifers or reservoirs
- Other environmentally sensitive areas or locations

The resulting consequences of an environmental impact are the identification of the environmental receptors that may be affected, and could also include the additional costs of:

- Emergency response
- Clean-up
- Restoration
- Long-term monitoring
- Fines
- Loss of reputation
- Permit withdraw or restrictions

3.2.5 Likelihood

Likelihood is the measure of the expected probability and frequency of an event occurring (i.e., a 1 percent chance, a frequency of 0.01 events/year). The factors that comprise the estimate of likelihood include:

- Event rate—estimate of the frequency of occurrence of the initiating event, which may be influenced by:
  - Mode of transportation
  - Route conditions (i.e., track and road condition)
  - Length of route
  - Speed
  - Specific carrier performance
  - Civil disorder, theft, etc. along the route
- Conditional spill probability—the probability that, given an event has occurred, a subsequent release will follow (e.g., release of a hazardous material due to an initiating accidental event)
- Release size probability distribution—the probability for release of a given size following an accident (for example, should the event occur, there is a 60 percent chance of a small leak, a 30 percent chance of a large puncture, and 10 percent chance of a catastrophic failure of the container)

Specific considerations for estimating the likelihood of an event involving different modes of transportation are presented in *Guidelines for Chemical Transportation Risk Analysis* (CCPS, 1995), and include the following transportation systems:

- Pipelines
- Rail
- Trucks
- Barges and other vessels traversing inland waterways
- Ocean-going vessels
- Intermodal

3.2.6 Differences in Safety and Security Risk Definitions

While security is presented separately in Chapter 6, many of the risk definitions and concepts for safety and security are similar. One difference is that security incidents are intentional, rather than accidental. This difference is the basis for understanding the hazards, potential consequences, and likelihood of a security scenario. Table 3.3 compares the basic elements of safety and security risk analysis for the listed risk factors. The potential range of incidents and the resulting consequences have similarities and differences. The incident in both cases is the release of hazardous material or energy from its transportation container from the accident for security event. The consequences are the potential impacts to people, property, the environment, or places of iconic value. The difference in a security analysis is that, on average, the incident magnitude (release size) may be much greater than an accidental release, since the intent is to maximize damage under the worse possible conditions. Additionally, if the hazardous material were stolen, it could be released—as is or in a weaponized form—in a location with a greater potential for consequences than envisioned along the transportation route itself.

Table 3.3 Comparison Safety and Security Risk Assessment

Risk Factors	Safety Assessment	Security Assessment
Initiating Event	• Accident • Non-Accident	• Direct Attack • Theft/Diversion • Tampering
Incident	• Release of hazardous material or energy	• Release of hazardous material or energy • Theft and weaponization of hazardous material
Consequence	• Potential impacts to people, property, and environment	• Potential localized impacts to people, property, environment, or systems/places of iconic value • Remote deployment of material that was stolen, focused on a target with higher potential consequences
Likelihood	• Based on historical accident data and understanding of safeguards in place and their reliability	• Based on threat, capabilities and vulnerability and understanding of safeguards in place and their reliability

The biggest differences between safety and security are the initiating events and the factors that make up the likelihood. For safety, the initiating events usually result from accidents or other non-accident causes. For security, however, the events are from a direct or opportunistic attack with the intent to release, steal, or contaminate the contents of a shipment.

3.3 RISK ANALYSIS PROTOCOL

Before a risk analysis is initiated, it is important to specify the scope and have a protocol in place for these activities. For a transportation network, the scope may need to consider:

- The entire supply chain
- A segment of the transportation network
- A single hazardous material or a mixed shipment
- A subset of all the materials shipped
- One mode or route of transport

The overall theme of this book is to put lower-level risks into the appropriate perspective and focus increased energy and resources on the most important or complex transportation issues. Figure 3.3 presents a risk assessment protocol that can be applied to accomplish this goal, with the first step being the identification and prioritization process.

Figure 3.3 Risk Assessment Protocol

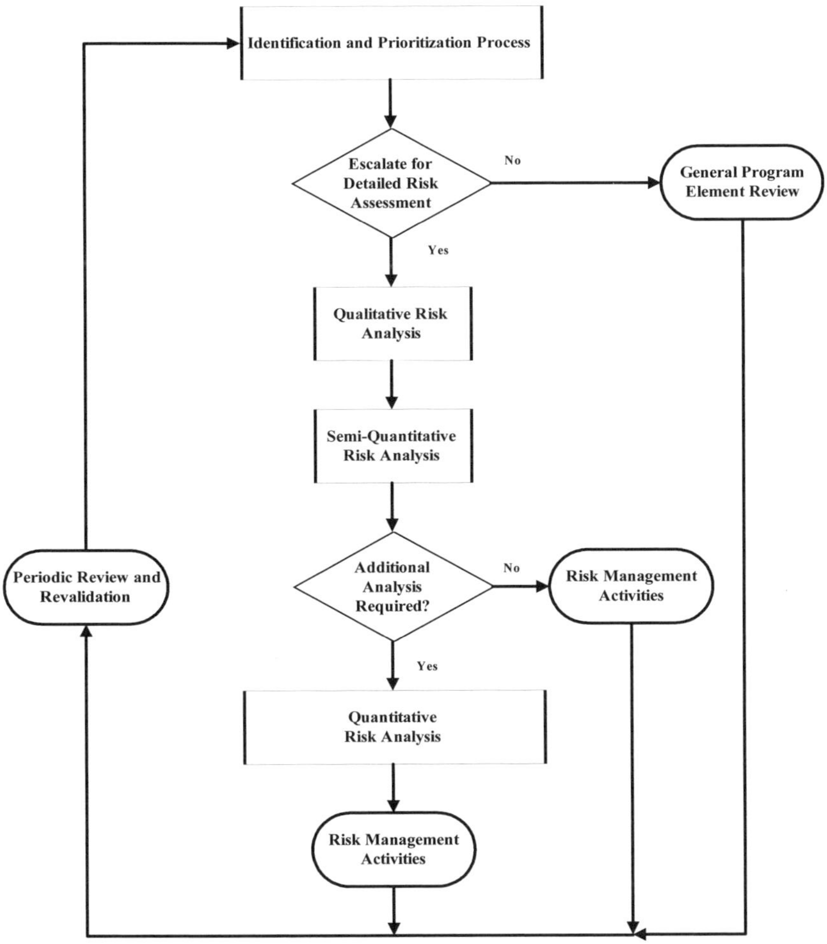

For those risks that require additional evaluation, many hazards may be identified, evaluated, and mitigated through use of more detailed analysis techniques. Qualitative/semi-quantitative risk analysis is presented in Chapter 4. In practice, the most risk analyses are qualitative/semi-quantitative in nature, and only lead to a more detailed quantitative analysis (Chapter 5) when a simpler approach cannot provide a clear understanding of the risk and more information is required for proper management. Additionally, qualitative analysis typically precedes all other higher levels of analysis as it is an essential data-gathering step of the risk analysis process.

The results of a risk analysis are fed into the risk reduction activities (Chapter 7) where different options for managing risk are evaluated and compared. For all issues that go through qualitative, semi-quantitative, or quantitative analysis, periodic re-evaluation of the level of risk should be conducted as part of the "sustainability" process described in Chapter 8, with changes implemented as required.

3.4 IDENTIFICATION AND PRIORITIZATION ACTIVITIES

The identification of hazardous material transportation scenarios stems from the identification and prioritization step of the TRM framework presented in Chapter 1. This process results in the generation of a list of activities and transportation scenarios that may require special attention, including additional resources for further risk analysis and risk management activities.

Identification and prioritization is used to rank chemicals, movements, and other factors for escalation, which may result in more detailed risk analysis. The components of a prioritization process can be included in a company-specific process and may consist of:

- Cataloging hazardous materials transported
- Documenting all modes of transport
- Identifying sensitive areas along the transit route
- Understanding interaction with other stakeholders in the supply chain

Appendix B, on the enclosed CD, contains a number of industry tools that are used by companies to conduct a risk prioritization. These tools are intended to provide examples, and can be modified as needed.

3.4.1 Cataloging Hazardous Materials and Modes of Transportation

The hazards of the chemicals that may be transported are discussed in regulations, defined by the hazard class, and presented in Section 3.2.1 of this chapter. In addition, there are other published lists of chemicals of concern from a worker safety, public safety, environmental, and security perspective. For example, a material may not be a regulated toxic, but may have certain characteristics such as chronic toxicity or reproductive effects that may warrant a more detailed risk evaluation. This step is not limited to chemical hazards, and may also include:

- Number of shipments
- Quantity per shipment
- Mode of transportation
- Container type

- Previous incidents
- Incident potential

3.4.2 Identifying Sensitive Areas Along the Transportation Route

The extent of the consequences of a transportation incident is dependent on the different sensitive areas and distance from the route to these areas. A sensitive area (or potential "receptor") that might be considered along a transportation route may include:

- Crossings of waterways or other environmentally sensitive areas
- Areas of high population density
- Nearby schools, hospitals, nursing homes, daycare centers, malls, prisons, or other population centers with large numbers of people
- Buildings, bridges, tunnels, and other infrastructure
- Areas with increased theft potential

3.4.3 Interactions with Other Stakeholders in the Supply Chain

A company's role in the supply chain, and the complexity of the network, can influence the need for further risk analysis. If a company is a small producer, as compared to others in the market, it may be beneficial to:

- Discuss the transportation of this material with the larger companies/carriers
- Compare safety risk management processes
- Benchmark potential transport scenarios
- Benchmark additional safety measures that may be applicable to your operation
- Implement specific safety changes to bring your operation into alignment with the performance level of the major shippers
- Identify emergency response capabilities and resources

While individual risk may appear manageable based on a review of hazards, mode of transportation, and typical operations, unexpected outcomes could occur if the full transportation network is not completely understood. Therefore, as part of any prioritization activities, the complexity of the transportation system should be reviewed to help ensure all potential issues are included. These complexities may include:

- Number of carriers
- Carrier safety record
- Carrier safety operating procedures
- Carrier compliance with applicable regulations
- Different modes of transportations
- Transfer operations and points along the route
- Different states, regions, and countries from origin to destination

3.4.4 Identification and Prioritization Summary

As detailed in this chapter, there are a number of transportation issues that can be reviewed alone or in combination to identify those requiring additional detailed analysis. The identification and prioritization step can be completed using a checklist to identify issues that may require escalation, including:

- Hazards of the materials being transported
- Specific initiating events that may be considered
- Potential consequences and impacts while in transit
- Accident history
- Potential gaps in safety practices
- Sensitive areas along the route for people, property, and the environment
- Complexity of the transportation activities and number of stakeholders in the supply chain

Once the issues along the supply chain are identified, there are a number of options available for identification and prioritization. As shown in Figure 3.4, these prioritization options include:

- *Hazard Based*—where all chemicals of a certain class are ranked the same (e.g., acute toxics, flammability rating, ecotoxics)
- *Consequence Based*—where potential high risk areas along the route are identified (e.g., high population density, environmentally sensitive areas, critical infrastructure)
- *Likelihood Based*—where recent history (internal and external accident performance of a specific carrier or of the industry as a whole) drives priority
- *Risk Based*—where the potential consequence and likelihood are used to prioritize

As illustrated in Figure 3.4, the different techniques can be used alone or in combination to escalate an issue requiring additional analysis. For example, a company may always require a detailed risk analysis for a specific chemical based solely on its hazard rating, whereas another company may only require additional analysis if the chemical is shipped by a specific mode and/or in large specified quantities.

The hazard-, consequence-, and likelihood-based approaches are founded on the potential issue or combination of issues. The risk-based approach discussed here is qualitative and could result in a subjective estimate of the level of risk (based on chemical hazards, potential impacts alone the route, and likelihood of an accident).

Regardless of the prioritization process utilized, the end result is a screening of transportation issues that can be limited to a general program element review, with a subset of issues escalated for more detailed analysis.

Figure 3.4 Identification and Prioritization Process

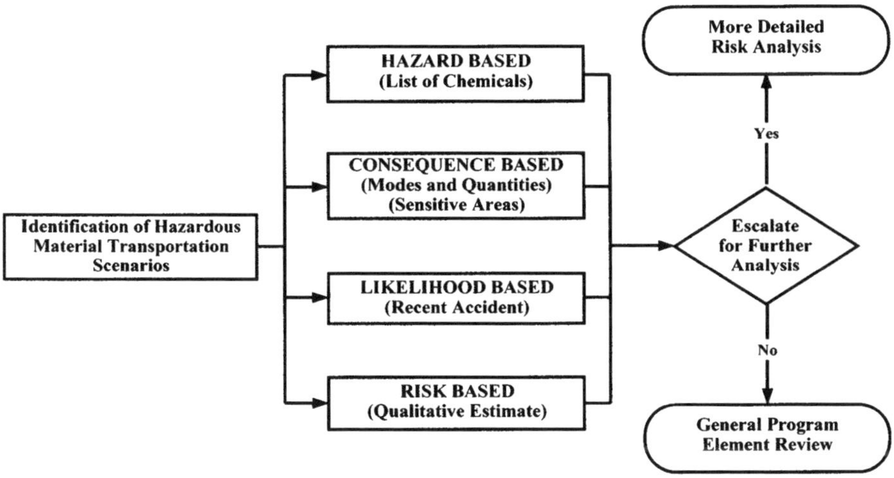

3.5 XYZ CHEMICAL EXAMPLE— IDENTIFICATION AND PRIORITIZATION

As discussed in Chapter 2, XYZ Chemical has brought in a new manager of risk management who has begun an initiative to develop a formalized global risk management program. While the company has a strong safety management program in the North America and Europe, the focus of the example is on ensuring that the foundational programs in the company's other regions of operation meet the corporate environmental, health, and safety (EHS) standards, as well as integrating all operations into a centralized risk management process. With the risk management system in place (Chapter 2), this organization continues with the next phase of the risk management program, the identification and prioritization process of chemicals and modes of transport that will require additional risk analysis.

While the identification and prioritization activity can be conducted at any level in the organization, this activity is being directed by corporate-level management and will include a corporate-wide view and additional follow-up actions for all hazardous material transportation activities. As detailed in this chapter, the prioritization step may include a review of any of the following to identify issues (or other considerations depending on the characteristics of the transportation operations):

- Hazards of the materials being transported
- Specific initiating events that may be considered
- Potential consequences and impacts while in transit

- Accident history
- Potential gaps in safety practices
- Sensitive areas along the route for people, property, and the environment
- Complexity of the transportation activities and number of stakeholders in the supply chain

Since this is a high-level corporate review at this point, the prioritization activity is limited to the chemicals, their physical properties, and the potential consequences and impact that could result from an accidental release while in transport to or from an XYZ Chemical plant. During the risk analyses presented in the continuing example in Chapters 4 and 5 (which focus on a specific facility's transportation activities), the other operational factors above are reviewed in greater detail. The prioritization process at this point, therefore, will be hazard- and consequence-based with additional detail on consequence, likelihood, and risk addressed for those chemicals and modes of transportation identified for escalation. Additionally, the review is limited to safety, since the focus of Chapter 6 is exclusively on security. In practice, safety and security can be completed together, but have been separated in this book and example for clarity.

XYZ Chemical has similar operations consisting primarily of pesticide and herbicide manufacturing operations in the United States, Europe, Asia, and South America. Since the operations are similar, the prioritization process (illustrated in Table 3.4) consists of cataloging all chemicals in transit (in and out of their facilities), modes of transport, quantities, and number of annual shipments. Specifically, the prioritization process is a combination of the hazards- and consequence-based prioritization processes illustrated in Figure 3.4. If this process were expanded to a likelihood- or risk-based process, then additional factors (e.g., the number of shipment, length of shipment) may have been included in the prioritization. With this data collected, the hazards of each chemical and the potential impacts are documented based on the chemical physical properties and quantities in transit. With all of the information collected, the following qualitative ranking is used to prioritize issues for escalation and identification of countries and operations that will need to conduct more detailed risk analyses:

- *Higher Risk*—Facilities whose transportation operations may pose a significant risk to people from an accidental release of toxics or flammables.
- *Medium Risk*—Facilities whose transportation operations may pose a moderate level of risk to people from an accidental release of toxics or flammables and/or a significant environmental impact potential.
- *Lower Risk*—Facilities whose transportation operations may pose a minor impact potential.

Table 3.4 Risk Priority Ranking—Worldwide Operations

Chemicals in Transit	Mode of Transit	Hazards	Potential Impacts	Risk Ranking
Chlorine	Railcar Tank Truck	Hazard Division 2.3 (toxic gas)	Toxic exposure to people located along transportation routes	**Higher**
Pesticides	Tank Truck	Hazard Class 6 (toxic substance)	Toxic exposure to people located along transportation routes	**Higher**
Herbicides	Truck	Hazard Class 6 (toxic substance)	Toxic exposure to people located along transportation routes and potential for environmental impact	*Medium*
Aqueous Acids	Railcar Tank Truck	Hazard Class 8 (corrosive)	Potential environmental impact	*Medium*
Ethylene Oxide	Railcar	Hazard Division 2.3 (toxic gas) Hazard Division 2.1 (flammable gas)	Potential for toxic exposure, vapor cloud fire, fire, or VCE	**Higher**
1,3-Butadiene	Railcar	Hazard Division 2.1 (flammable gas)	Potential VCE and fire hazard	*Medium*
Acrylonitrile	Railcar	Hazard Class 3 (flammable liquid)	Potential VCE and fire hazard	*Medium*
Aqueous Ammonia	Tank Truck	Hazard Class 8 (corrosive)	Potential environmental impact	Lower
Methyl Bromide	Truck (cylinders)	Hazard Division 2.3 (toxic gas)	Potential toxic release hazard	Lower
Formaldehyde Solution	Tank Truck	Hazard Class 8 (corrosive)	Potential environmental impact	Lower

The results of the prioritization process are detailed in Table 3.5. Three activities have been ranked as "Lower Risk" and will be dropped from further consideration. These chemicals have been ranked as lower risk because the concentration of aqueous ammonia and formaldehyde in combination with a small shipment volume presents only a minor environmental impact potential. Additionally, methyl bromide is used at some facilities, but it is received in small cylinders that have only a localized impact potential if released while in transit. These chemicals and modes of transport will, however, continue to be reviewed and monitored as part of the global risk management program, including a primary management system review. The remaining chemical and modes of transport will be escalated for qualitative risk analysis.

Table 3.5 Risk Priority Results—Worldwide Operations

Chemicals in Transit	Mode of Transit	Prioritization Risk Ranking	Post-Prioritization Activity
Chlorine	Railcar Tank Truck	**Higher**	Escalation – Qualitative Risk Analysis
Pesticides	Tank Truck	**Higher**	Escalation – Qualitative Risk Analysis
Herbicides	Truck	*Medium*	Escalation – Qualitative Risk Analysis
Aqueous Acids	Railcar Tank Truck	*Medium*	Escalation – Qualitative Risk Analysis
Ethylene Oxide	Railcar	**Higher**	Escalation – Qualitative Risk Analysis
1,3-Butadiene	Railcar	*Medium*	Escalation – Qualitative Risk Analysis
Acrylonitrile	Railcar	*Medium*	Escalation – Qualitative Risk Analysis
Aqueous Ammonia	Tank Truck	Lower	Primary management system review
Methyl Bromide	Truck (cylinders)	Lower	Primary management system review
Formaldehyde Solution	Tank Truck	Lower	Primary management system review

Based on these initial findings, qualitative operational-level reviews of facilities transporting any of the seven medium- or high-risk chemicals were directed from corporate. The goal of these risk analyses is to further understand XYZ Chemical's transportation risks and to identify where risk reduction options may need to be considered. This transition from a global to operational-level review will be the focus as the example continues to evolve in Chapters 4 and 5.

REFERENCES

Center for Chemical Process Safety, *Guidelines for Analyzing and Managing the Security Vulnerabilities of Fixed Chemical Sites*, American Institute of Chemical Engineers, New York, 2003

Center for Chemical Process Safety, *Guidelines for Chemical Process Quantitative Risk Analysis, Second Edition,* American Institute of Chemical Engineers, New York, 2000

Center for Chemical Process Safety, *Guidelines for Chemical Transportation Risk Analysis*, American Institute of Chemical Engineers, New York, 1995

Center for Chemical Process Safety, *Guidelines for Implementing Process Safety Management Systems*, American Institute of Chemical Engineers, New York, 1994

Lees, F.P., *Loss Prevention in the Process Industries, Second Edition,* Butterworth Heinemann, London, 1996

4
QUALITATIVE AND SEMI-QUANTITATIVE RISK ANALYSIS

The risk assessment protocol outlined in the previous chapter suggests an approach for a systematic analysis and evaluation of risk, by first identifying potential "higher-priority risk" transportation activities through a prioritization process. Those issues determined to be lower priority, or "lower risk," are managed as part of an overall management system (Chapter 2) with the higher-priority items escalated for a more detailed safety and/or security analysis. The goal for the supply chain is a management plan appropriate for the level of risk, resulting in a rational balance of safety and security.

The next three chapters present techniques for analyzing safety (Chapters 4 and 5) and security (Chapter 6) issues concerning the transportation of hazardous materials. Risk analyses can vary from simple, screening-level analyses intended to provide an estimate of the types and levels of risk to very comprehensive, detailed analyses of incident probabilities, hazardous materials release probabilities, and modeling of the adverse effects resulting from an accidental hazardous material release. Typically, a phased or tiered approach to analyzing risks, which progresses in complexity as warranted, makes the most efficient use of resources. As the level of complexity is increased, the analysis builds on the information from the earlier analyses. This continued data collection and increasingly detailed level of risk analysis result in a better understanding of the transportation operation, and feeds into the decision-making process that ultimately determines if additional resources will be needed to manage risk.

This chapter focuses on qualitative and semi-quantitative techniques that can be used to analyze the safe transport of hazardous materials. These simple and efficient risk assessment approaches represent the level of complexity needed to sufficiently understand and make decisions for the majority of hazardous material issues. A fully quantitative approach (Chapter 5) may be needed if the level of detail developed with qualitative or semi-quantitative approaches is inadequate to confidently make final risk management decisions. The choice of a more quantitative approach also may be needed if the cost of the identified risk mitigation options is high, warranting a more detailed understanding of the operational risks and risk reduction benefits. The evaluation and strategies for selecting risk reduction options is the topic of Chapter 7.

4.1 OVERVIEW OF QUALITATIVE AND SEMI-QUANTITATIVE RISK ASSESSMENTS

Figure 4.1 illustrates the risk assessment process for qualitative and semi-quantitative techniques that can be applied to transportation safety scenarios. Typically, these will be a scenario and issues developed and escalated for more detailed analysis during the identification and prioritization process described in Chapter 3.

As shown in Figure 4.1, issues escalated from the identification and prioritization process are first evaluated using qualitative techniques. A qualitative risk analysis is based primarily on description and comparison using historical experience and engineering judgment, with little quantification of the hazards, consequences, likelihood, or level of risk. This first level of risk analysis should be typically conducted before further semi-quantitative or full quantitative evaluations are initiated. Even at this stage of analysis, a reasonable risk assessment decision can be reached. With the findings documented, the activities managed as part of the transportation system, and the focus moved to other more complex issues escalated through this process. The qualitative risk analysis step also collects additional information, building from the prioritization process, which will be necessary for the more detailed analyses. This data collection activity, and ability to better understand the operation, is another reason for not bypassing qualitative risk analysis in lieu of more detailed risk analysis approaches.

For those issues or transportation scenarios requiring more detail and insight than a qualitative approach offers, the next level of analysis is a semi-quantitative evaluation. A semi-quantitative risk analysis includes some degree of quantification of consequence, likelihood, and/or risk level. Once evaluated, through the combination of consequence and likelihood, a risk-based decision can be made from the analysis or additional refinement in consequence or likelihood may be necessary. This refinement of the risk analysis may rely on additional information included as part of a semi-quantitative analysis or, as warranted, on some minimal level of quantitative techniques described in the next chapter.

Regardless of the risk analysis technique utilized, risk decisions and evaluation of risk mitigation options can only be accomplished if evaluation criteria are defined. Evaluation criteria are the metrics used to determine if additional risk controls will be needed as well as the means to compare the benefits of competing risk mitigation options. While the selection of company-specific evaluation criteria is beyond the scope of this guideline, it is critical that risk analyses be evaluated against metrics consistent with the risk tolerance of the individual companies involved in the supply chain activity being reviewed. Ideally, risk evaluation criteria and the risk management philosophy should be applied consistently for all company risk decisions.

Figure 4.1 Qualitative and Semi-Quantitative Risk Assessment Process

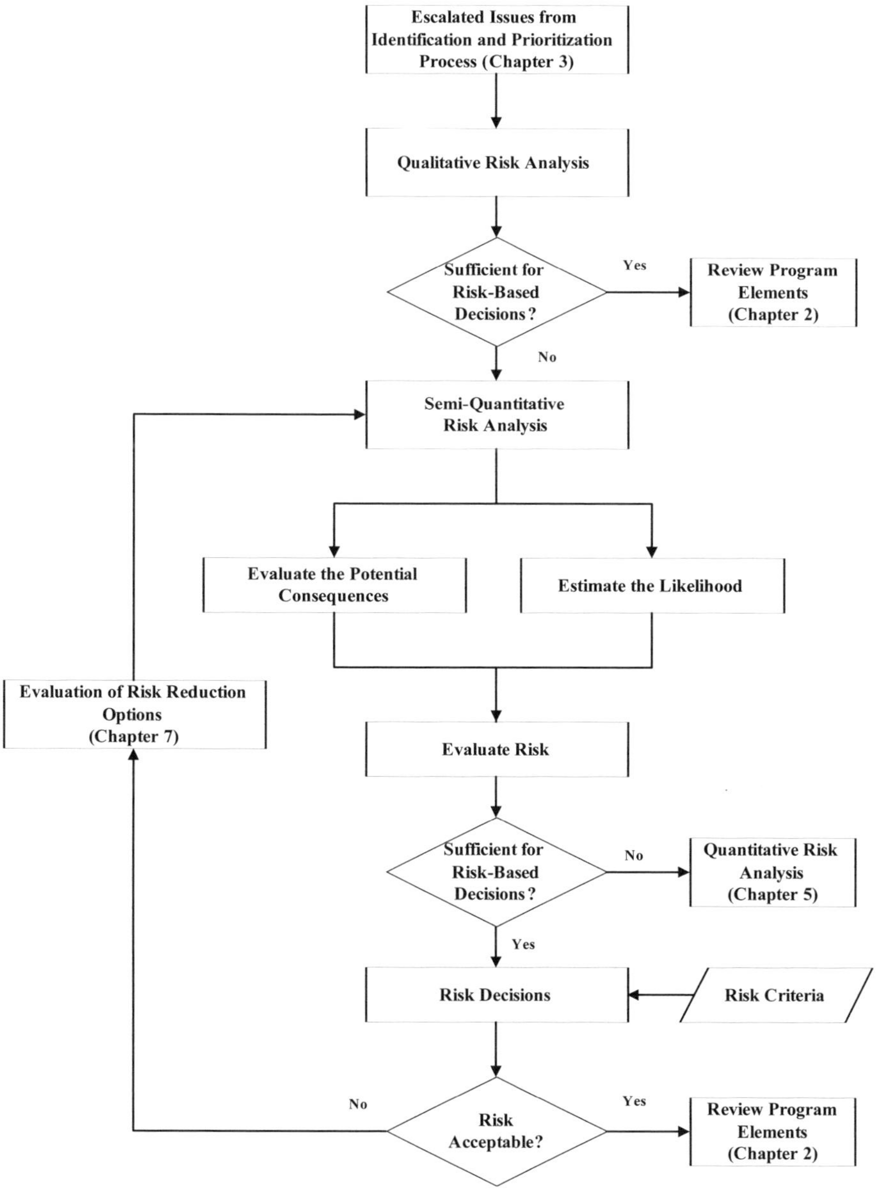

4.2 QUALITATIVE RISK ANALYSIS

Qualitative risk analyses can vary in scope depending on the complexity of the transportation activity being evaluated. Depending on the issues identified during the prioritization process, qualitative risk analyses can range from an evaluation of a chemical, movement, segment of the route, or the movements across the entire supply chain. The application of qualitative risk analyses is discussed below and often entails:

- Establishing benchmark comparisons
- Identifying and elevating issues of concern
- Understanding the impact of anticipated changes

4.2.1 Establishing Benchmarking Comparisons

Benchmarking is the comparison of current operating practices to internal or external company practices, industry best practices, and regulatory standards. Any differences identified through a benchmarking activity can be used to develop recommended changes in operating practices so a company, carrier, or others in the supply chain can be aligned with, or exceed, the desired industry practices and/or regulatory requirements.

A company involved in the transport of hazardous materials may, or may not, be the industry leader in the risk management of a specific chemical's transport, certain mode of transport, or specific activity (e.g., emergency response). Therefore, instead of conducting an entire risk analysis for these types of activities, benchmarking enables a company to collect, review, understand, and align itself (as necessary) with the leader's practices for these types of movements.

In addition to benchmarking focused on operating practices, other specifics of the route may be compared using benchmarking, including:

- Miles transported
- Carrier accident rates
- Proximity to sensitive receptors
- Recent incidents

These types of benchmarking issues can be used to identify other issues that may require further analysis due to the potential for severe consequences or higher perceived likelihood for an accident. The issues identified through a qualitative analysis can be directly used as inputs in a semi-quantitative analysis with a focus on developing a better understanding of these issues and their influence on the level of risk, and need for additional risk mitigation measures.

Table 4.1 provides some examples of benchmarking opportunities to identify operational changes that may reduce transportation risk or to identify and elevate issues for a more detailed risk analysis. The list in Table 4.1 is not intended to be an exhaustive list of options and, therefore, should be customized to the specific needs of the transportation operation.

Table 4.1 Examples of Benchmarking Transportation Risk Activities

Benchmark	Item of Interest
Chemical Hazards	• Material classification and packaging – Hazard class – Packing group – Transportation Requirements/Restrictions
Industry Experience	• Associations – Industry – Trade – Other standards/guidance organizations • Incident Reports
Container Design and Operating Practices	• Equipment details • Performance analyses • Equipment maintenance • Quality assurance • Loading/unloading procedures • Control systems
Safety & Security	• Product stewardship – Customer review/approval – Carrier product seminars – Equipment repair – Equipment cleaning • Inspection practices • Route analyses and selection • Shipment monitoring • Security program and practices • Carrier selection – Evaluation and periodic review – Training – Safety initiatives – Security programs and practices • Emergency response and preparedness – Resources – Training and qualifications – Drills and exercises

A simple example of a benchmarking exercise might be a review of how the various business units in a company ship flammable gases [Hazard Class] via rail [Mode or Size of Shipment]. By using these parameters to set the benchmark comparison, some or all of the other parameters can be compared and contrasted such as:

- Are formal compliance assessments conducted [Audits]?
- Are specific design standards and practices employed consistently [Design Standards and Practices]?

- What is the range and depth of pre-load and pre-ship inspection efforts [Inspection Practices]?
- What is the nature of customer and supply chain partner interaction to assure the safe transport and handling of the materials [Product Stewardship]?
- Are security practices consistent [Shipment Monitoring]?
- Are external benchmarking activities conducted [Standards/Guidelines]?

Without a very detailed analysis, benchmarking can result in quick and effective comparisons, with valuable learnings.

In the descriptive example above, the benchmarking was internal to one particular company where it would be relatively easy to access all the required information to complete the analysis. Benchmarking becomes more difficult if data from other companies or organizations need to be collected, and it becomes much more important to clearly define the information requested. Information requests that are unclear or open to multiple interpretations can lead to the collection of flawed data that cannot be compared and may ultimately result in erroneous benchmarking results and misinformed risk decisions. Also, if data are collected via a one-time survey, unlike benchmarking within one organization, the ability to seek clarification is significantly reduced, if not impossible. Thus, the design and execution of benchmarking require careful planning and design. Additionally, as part of the design and documentation, care should be taken to avoid suggestion that any company's program or activities are inadequate or deficient. Likewise, if published data will be used to support a benchmarking study, care is needed to assure that the basis of the data used from different sources is comparable and consistent.

4.2.2 Identifying Issues for Further Evaluation

Table 4.2 provides a checklist of some of the information that can be collected as part of a qualitative risk analysis. This checklist is intended to be a starting point for identifying issues that may require further evaluation. It is included in Appendix B, on the enclosed CD, and should be altered, shortened, or expanded to meet the needs of your particular operation.

Table 4.2 Example Checklist of General Review of Transportation Issues

NO.	ISSUE	RESPONSE
1	What material is being (or to be) transported?	
2	What are the physical and chemical properties of this material?	
3	Under what physical conditions (temperature and pressure) will this material be transported?	
4	Can the release of this material cause harm to people in the immediate vicinity or elsewhere?	
5	What other materials will be stored in close proximity during shipment?	
6	Could these different materials react during spillage to cause an emergency?	

NO.	ISSUE	RESPONSE
7	What is the proposed mode of transportation?	
8	What size and type of containers will be used?	
9	What is the total size of the shipment?	
10	How many shipment loads are proposed?	
11	How will the containers be secured?	
12	How will the containers be labeled?	
13	Who will load the shipment?	
14	What size and type of carrier vehicle will be used?	
15	How will the carrier vehicle be inspected?	
16	Who will operate the carrier vehicle?	
17	Are there any special driving precautions required while transporting this material?	
18	Are there any hazardous warning signs on the carrier vehicle advising other traffic to stay clear?	
19	What is the destination of the shipment?	
20	Are there any specific instructions for accessing the destination facility?	
21	Is the off-loading of the shipment carried out by trained workers?	
22	When is the shipment required?	
23	What is the proposed route?	
24	What alternate route is available in the event of traffic or other restrictions?	
25	Does the route pass through any residential or commercial areas?	
26	Is the proposed route heavily traveled by others?	
27	Does the proposed route include travel through tunnels?	
28	Does traffic along the proposed route vary significantly over a 24-hour cycle?	
29	Will the proposed shipment be transported during peak traffic periods?	
30	Are there any load restrictions along the proposed route?	
31	Is the carrier operator qualified to deal with minor spillage or loss of containment incidents?	
32	Could the material container easily fail or rupture during a collision or upset?	
33	In a major collision accident, is there adequate warning time to evacuate people to a safe location?	
34	Following a major spill or release, can the area be adequately cleaned to reduce the exposure risk to the public?	
35	Are there any special safeguards that will be or should be implemented to address the concerns noted above?	

4.2.3 Understanding the Impact of Anticipated Changes

Qualitative reviews should identify any concerns associated with a change to current operating practices. Such changes may include:

- New or revised regulations
- New material being shipped
- New mode of transportation
- New shipping containers or other equipment (including materials of construction for all components—even gaskets)
- New transportation conditions (i.e., temperature or pressure)
- New supplier
- Increased quantity shipped
- Increased shipments
- Distribution to new states, regions, or countries
- Change in routing
- Change of carrier
- Changes in personnel
- Any other change in business strategy/importance

Any of the qualitative analysis techniques can be applied with the selection based on the change being considered. Again, following the risk analysis protocol, issues that cannot be resolved using qualitative techniques should be escalated for a more detailed risk analysis starting with the semi-quantitative analysis technique described in Section 4.3.

4.2.4 XYZ Chemical Example—Qualitative Risk Analysis

In Chapter 3, XYZ Chemical conducted a corporate-level identification and prioritization of all worldwide hazardous material transportation activities. Based on this hazard-based prioritization activity, the chemicals listed in Table 4.3 were identified as potentially "higher risk" and escalated for further facility/operational-level reviews. Facility review means the inbound and outbound hazardous material transportation activities for each XYZ Chemical plant moving these chemicals.

At this point in the example, the company-wide reviews completed in Chapters 2 and 3 are replaced with a first-level qualitative risk analysis focused on the transportation activities of a single XYZ Chemical facility. To continue the example, a single XYZ Chemical facility is selected. This facility will continue to be evaluated as the level of analysis detail increases from qualitative to semi-quantitative in this chapter to quantitative risk analysis (Chapter 5). As the level of detail increases, the analysis will be directed at specific questions that remain following each level of analysis. This facility will continue to be the focus of Chapter 6 where the security vulnerability of the hazardous materials in transit is evaluated, Chapter 7 where risk reduction options are evaluated, and Chapter 8 where the ongoing management of risk continues in the future.

The specific facility, illustrated in Figure 4.2, is located in XYZ Chemical's Asian market, and receives and ships all seven of the chemicals of concern identified in the corporate-level prioritization activity. Specifically for this facility, these chemicals include:

- Raw materials
  - Chlorine and ethylene oxide in railcar quantities
  - Butadiene and acrylonitrile in tank trucks
- Products
  - Repackaged chlorine in 150-pound cylinders and ton containers
  - Liquid pesticide in both bulk and non-bulk quantities
  - Non-bulk herbicide (powder form) shipped in distribution packaging via truck
- Wastes
  - 32% hydrochloric acid (HCl) shipped out in railcars

Table 4.3 XYZ Chemical Prioritization Process Results

Chemicals in Transit	Hazards	Potential Impacts	Risk Ranking
Chlorine	Hazard Division 2.3 (toxic gas)	Toxic exposure to people located along transportation routes	**High**
Pesticides	Hazard Class 6 (toxic substance)	Toxic exposure to people located along transportation routes	**High**
Ethylene Oxide	Hazard Division 2.3 (toxic gas) Hazard Division 2.1 (flammable gas)	Potential for toxic exposure, vapor cloud fire, fire, or VCE	**High**
Herbicides	Hazard Class 6 (toxic substance)	Toxic exposure to people located along transportation routes	Medium
Aqueous Acids	Hazard Class 8 (corrosive)	Potential environmental impact	Medium
1,3-Butadiene	Hazard Division 2.1 (flammable gas)	Potential VCE and fire hazard	Medium
Acrylonitrile	Hazard Class 3 (flammable liquid)	Potential VCE and fire hazard	Medium

Figure 4.2 XYZ Chemical Facility

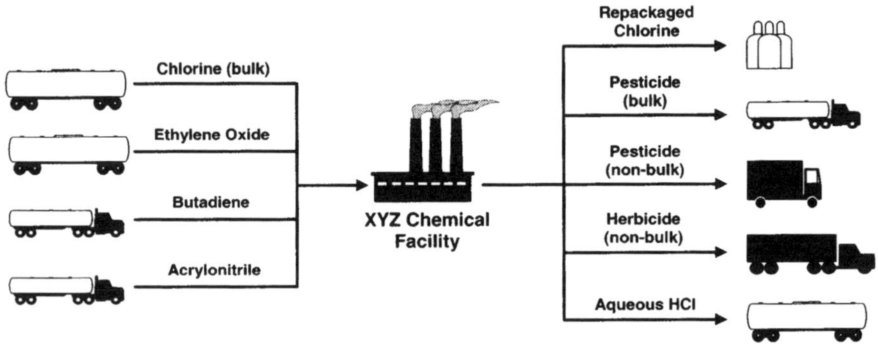

The site-level health and safety manager was given notice from corporate for the need to conduct the qualitative risk analysis of the plant's hazardous material transportation operations. The safety manager was provided with a questionnaire regarding all chemicals of concern, shipping quantities, modes of transit, and the route characteristics (similar to the checklist found in Table 4.2). This information was transmitted back to corporate for analysis, which included a corporate-level comparison to the other XYZ Chemical facilities' risk results. Based on the information collected at the site level, the following was determined and reviewed by corporate for this facility:

- Modes and quantities
  - Chlorine in bulk rail cars pose a potential for a large accidental consequence zone (regardless of transportation route) and has a much higher consequence potential than repackaged chlorine in 150 pound cylinders or ton containers.
  - Transport of pesticides (both bulk and non-bulk) pose a potential for a consequence of concern if the route passes near areas of high population density.
  - Herbicide in powder form poses primarily an environmental concern.
  - Hydrochloric acid (32% aqueous solution) poses primarily an environmental concern.
  - Ethylene oxide poses a potentially significant consequence if transported near areas of high-population density or near critical infrastructure along the route.
  - Butadiene and acrylonitrile pose a moderate consequence if transported through high-population centers or critical infrastructure along the route.
- Sensitive areas along the route
  - Inbound chlorine does not pass through or near any major cities.

  - Repackaged chlorine is shipped via truck into areas of moderate to high population.
  - Both bulk and non-bulk pesticides are shipped through areas of high population density.
  - The herbicide route does not pass near any significant environmental receptors.
  - Hydrochloric acid (32% aqueous solution) is not shipped near areas of high-population density or near any significant environmental receptors.
  - Ethylene oxide is shipped through areas of moderate population density.
  - Butadiene and acrylonitrile are only shipped through rural areas.
- Recent incidents
  - There had been a number of accidents involving bulk pesticide shipments in Asia and South America.

Based on the combination of potential consequence and route characteristics, the following activities were screened out by corporate for this specific facility:

- Herbicide product shipments
- Repackaged chlorine cylinder transportation
- Waste stream of 32% HCl
- Butadiene and acrylonitrile tank truck receipts

From this qualitative review of the operation, it was recommended that certain activities undergo a more detailed risk analysis. The analysis will be facilitated by a corporate representative and performed by a team of local personnel and other supply chain partners, as required. The activities that will be included in this more detailed review will include:

- Bulk chlorine in railcars
- Repackaged chlorine in ton containers
- Bulk and non-bulk pesticide shipments
- Ethylene oxide railcar deliveries

As part of the corporate-level review, a benchmarking of the bulk chlorine railcar operations across the regions was conducted. From the benchmarking activity, XYZ Chemical identified the need to standardize a program for chlorine shipments. The program was developed based on the company's European operations, since this is the region where the company has its largest annual chlorine movements and the most years of experience. Therefore, a program for chlorine railcar shipments was rolled out worldwide and included specifics such as railcar design and construction, alignment with UN Model Regulations and RID in all developing counties with no defined dangerous goods regulations, and a rigorous carrier qualification program. With the implementation of this chlorine railcar program, the risk of the raw material deliveries to the Asian plant was reduced and screened out for further analysis.

Table 4.4 summaries the results of the qualitative risk analysis. Three of the raw materials operations (chlorine, butadiene, and acrylonitrile) and three of the products/wastes (chlorine cylinders, herbicides, and aqueous HCl) were screened out for further risk analysis. Therefore, four chemical movements were escalated for a more detailed semi-quantitative risk analysis

Table 4.4 Qualitative Risk Results

Chemical	Mode	Qualitative Risk Ranking	Action
Chlorine	Railcar	**Higher**	Benchmarking (Standardized Program)
Chlorine	Cylinders	Lower	Screened Out
Chlorine	Ton Container	**Higher**	**Escalated**
Ethylene Oxide	Railcar	**Higher**	**Escalated**
1,3-Butadiene	Tank Truck	Lower	Screened Out
Acrylonitrile	Tank Truck	Lower	Screened Out
Pesticides (bulk)	Tank Truck	**Higher**	**Escalated**
Pesticides (non-bulk)	Truck	**Higher**	**Escalated**
Herbicides	Tank Truck	Lower	Screened Out
Aqueous HCl	Railcar	Lower	Screened Out

4.3 SEMI-QUANTITATIVE RISK ANALYSIS

A qualitative analysis may result in an escalation of specific issues that require a more detailed analysis before appropriate decisions can be made regarding the level of risk and suitability of potential risk mitigation options. The next level of risk analysis is semi-quantitative. Semi-quantitative analyses differ from qualitative analyses in that some degree of quantification is conducted to refine the estimates of consequence, likelihood, or level of risk.

One of the major benefits of a semi-quantitative risk analysis is that the technique can be applied and results understood by a wide range of stakeholders in the transportation field. Unlike full quantitative approaches, these types of risk analyses do not require specialized risk management experts. Even with the ease of application, however, personnel involved in these activities need to be knowledgeable in the operations under evaluation and the use of semi-quantitative methods. To assure proper outcomes from these risk management activities, the team assembled for a semi-quantitative assessment should be comprised of individuals with extensive transportation expertise, including safety and risk management, regulations, and logistics. Additional team members should be added as needed to address specific issues such as consequence modeling, appropriate emergency response procedures for the product characteristics, local routing at origin, and destination plant sites.

Some additional benefits of semi-quantitative techniques include:

- Wider availability and ease of use to the full range of hazardous materials supply chain personnel
- Ability to be applied relatively quickly and easily
- Ability to screen a number of issues, identifying the most critical that may require a more detailed review
- Ability to update previous qualitative risk analyses with additional information
- Efficient use of resources
- Ability to compare different transportation options
- Ability to easily make adjustments reflecting changing company priorities
- Ability to easily make adjustments to reflect changing transportation regulations
- Applicability to a wide range of transportation scenarios
- Ability to account for many factors that influence risk
- Ability to address a range of consequences and likelihoods
- Ability to screen out and identify those issues that may need to be escalated for further analysis
- Ability to provide a preliminary relative risk estimate

Certain factors may require some degree of weighting or quantification in a risk analysis. Table 4.5 presents factors associated with the consequences and likelihood that may need to be considered in a semi-quantitative risk analysis.

The factors listed under consequence in Table 4.5 can be inserted into a consequence model to determine the extent of vapor dispersion, thermal radiation from a fire, or overpressure from an explosion. There are a number of consequence models that can be utilized for these types of calculations ranging from order of magnitude estimates, simple calculations, free government-sponsored software, and proprietary models. The purpose of this guideline is not to review or recommend consequence analysis methods or tools. If more information is desired on consequence modeling, the *Guidelines for Chemical Process Quantitative Risk Analysis* (CCPS, 2000) should be reviewed. Frequently, however, at this stage of analysis it is often desirable to use very conservative professional-judgment types of estimates of consequences with more refined modeling approaches used as needed in subsequent iterations of semi-quantitative analyses. It is acceptable to have several semi-quantitative iterations prior to escalating to a quantitative review.

Table 4.5 Risk Factors

Risk Parameter	Factor Description/Influence
Consequence	Release type(s): Different accident and non-accident related incidents can be developed to represent the potential route scenarios. Release size: Full range of release sizes (e.g., leaks, large spills, catastrophic releases) can be evaluated to provide a range of potential outcomes that could occur during transit. Mode/container size: The quantity of material that can be released is dependent on the size of the shipping container, which in turn is dictated by the mode of transport. Evaluating the container size enables a comparison of the extent of the potential consequences between different modes as well as different transportation amounts for the same mode. Material conditions: The temperature and pressure of the contents in transit can greatly influence the potential extent of a chemical release. In general, higher temperatures and pressures result in more severe consequences. Range of potential consequences: Depending on the scenario, hazards of the material, the material conditions, and the location, the consequences of a release can range from a liquid spill, vapor dispersion downwind, immediate ignition and localized fire, delayed ignition resulting in a flash fire or explosion, or a BLEVE. Meteorological conditions: The ambient temperature, humidity, wind speed, and stability class can influence the potential downwind extent of a chemical release. In general, the lower the wind speed and more stable the atmospheric conditions, the longer the downwind dispersion due to the slow mixing and dilution with air. Evaluating the full range of potential weather conditions enables a better understanding of the potential consequences zones. Sensitive receptors along the route: The location of people, property, and other environmental receptors along the route, as well as their distance, enables the development of the potential impacts (fatality, injury, economic, environmental).
Likelihood	Accident rate: For most modes of transport the accident rate is reported in accidents per mile of transit (i.e., truck, rail, marine, air). For a pipeline, the accident rate is dependent on the size, material of construction, and the location of the pipeline (above or below ground, subsea), as well as other factors such as maintenance and external impacts (e.g., digging). Miles per transit: Fundamental unit of measure for rail, truck, marine, and air enabling industry, company, or carrier accident rates to calculate a predicted number of accidents per transit. Trips per year: Combining the number of trips with the accidents per transit will result in a predicted number of accidents per year. Release probability: Data can be developed specific to the route conditions (e.g., road/rail conditions, speed) and the type of accident (e.g., collision, derailment, external impact type, ship grounding) that can be used to develop a range of probabilities, and the size, of a release. Route characteristics: Include the crossing of environmental areas, passing of potential infrastructure, and distance location and density of people along the route. Meteorological conditions: The conditions affect the consequence zones, but probabilities can also be developed based on the frequency of the wind direction, which may vary from night to day, and which could increase or decrease the potential for a vapor cloud to travel toward a high population density along the route.

As will be discussed in more detail in Chapter 5, care must be taken when quantifying likelihood and consequence estimates. Issues and pitfalls may include:

- Comparison of different sources of data (specific accident data for a mode as compared to generic data)
- Variations in data quality
- Confidence in data sources
- Use of data from different transportation modes

It is strongly recommended that clear guidelines be developed for use when quantifying likelihood and consequence. Examples are presented later in this chapter.

4.3.1 Refinement of Consequence and Likelihood Estimates and Preliminary Mitigation Considerations

While qualitative techniques represent a very basic analysis, with little modification, they can be expanded and evaluated in a semi-quantitative risk analysis. Additionally, events and scenarios elevated for further analysis can be further evolved by considering the range of potential consequences (small to large) and/or likelihood (low to high probability of occurrence).

Several parameters may be reviewed as part of a semi-quantitative analysis, including the scope of the analysis, the potential consequences and range of consequences, and the likelihood of a transportation accident and subsequent chemical release. The various inputs to these components of a semi-quantitative analysis are detailed below.

Scope of Analysis

- Will the analysis include both accidents and non-accidents or be limited to one or the other?
- Will the analysis include the entire route or focus on specific chemicals, modes, or segments of the route?

Consequence

<u>Material Properties</u>

- What are the details of the hazards of the material as currently shipped?
  - toxicity
  - vapor pressure
  - polymerization potential
  - flammability
  - reactivity (water, other materials)
  - multiple hazards
- Can a less toxic or harmful material be shipped (i.e., solid vs. liquid, solution vs. anhydrous, intermediate vs. raw material)?

- Can the toxicity or harmful effects be reduced through environmental control (i.e., temperature) or chemical additives?

Populations (Receptors)

- Are there major population centers or sensitive ecosystems (including drinking water) along the route?
- Is housing along the route enclosed or open air (i.e., would the building structure provide some protection against fumes)?
- Is it possible to specify routing to avoid high-population or sensitive areas *without compromising road or rail quality* or drastically increasing the number of miles?

Emergency Response

- Are trained emergency response teams (private or public) available?
  - To what level are they trained?
  - How quickly can they respond?
  - How practical is it to expect the release to be contained?
- Are mutual aid response agreements in place with other manufacturers?
- What special equipment is required to mitigate a leak of the material (special pumps, sorbents, etc.)?
  - Who has this equipment?
  - In case of a transload, are appropriate transfer containers available?
- Are community notification/evacuation plans in place?
- Have joint drills been held with carriers, public responders, and communities along the route?

Likelihood

Mode of Transport

- Have alternative modes by number of trips, number of miles per trip, and accident frequency per mile been evaluated?
- What modes (rail, highway, air, bulk marine) are available within the region (and possible for the given origin-destination combination)?
- Could the number of trips be reduced by shipping via another mode?
- Might this reduce the risk in terms of frequency and consequence

Many other factors will depend upon the choice of mode, including carrier selection, equipment design, and routing.

Carrier Selection

- Have alternative carriers by the appropriate protocol (Shippers, Responsible Care®, or equivalent protocols) been assessed?
- Is the carrier included in the shipper's approved carrier program?
- What is the accident rate for this carrier and how does it compare to other hazardous material carriers?
- Does the carrier have an established safety management program?
- Does the carrier have an established security program, including background checks and route security procedures?
- When safety or security incidents have occurred, does the carrier investigate and take corrective action?

  Trucking-specific—Driver Performance

  - How does the carrier select drivers (length of experience, safe driving record, etc.)?
  - Have the drivers received training in hazardous material transportation and rules?
  - Can the drivers read and write (for understanding of hazardous bills of lading)?
  - Will specific (dedicated) drivers be used for this product?
  - Have the drivers been trained about the hazards of this product?

  Trucking-specific—Equipment Maintenance

  - Does the carrier have a preventive maintenance program for the tractors and other carrier-owned equipment?
  - What is the general age/condition of the equipment being used in this service? Are all mandatory inspections up-to-date?
  - Have breakdowns caused delivery disruptions?
  - Are adequate chassis available for ISO containers (to lower center of gravity), if applicable?

Equipment Design

- Have alternative choices for container design (including the minimum requirements specified by U.S. DOT or UN specification) been evaluated?
- Does the container meet a U.S. DOT or UN specification?
- Based on the hazard of the product, can/should the accident performance be improved by:
  - Increased shell/body thickness?
  - Insulation?
  - Head protection?

  - Other reinforcement?
  - Elimination of bottom outlets?
  - Improvement of valve protection?
  - Change in construction material?
- Is there a maintenance manual for the containers and has it been provided to the maintenance shops and reviewed with them?
- Are certain repairs witnessed by quality assurance inspectors (valve replacement, welding)?
- Are maintenance shops audited for compliance with these procedures?
- Are procedures in place to ensure that the containers are dry and not contaminated when returned from shop?
- Do written procedures (including torque specifications) exist for closure of container?
  - Have these been transmitted to loaders, customers, maintenance shops, etc.?
  - Are these procedures being followed?
- Is blocking and bracing sufficient for transport mode and route?
- Are ISO tanks "locked" onto chassis?

<u>Transport Considerations</u>

- Are there specific temperature/pressure conditions that must be maintained for material to be delivered safely?
- Is there a maximum number of days that the material can remain in transit (time-sensitive shipments)?
- Are sufficient procedures (e.g., tracking systems, communication with carriers and receivers, emergency options in case of delay or lost container) in place to ensure these conditions are met?

<u>Transport Speed</u>

- Are there carrier-imposed speed limits in effect?
- What is the maximum speed allowable?
- Do the limits vary along the routes of interest?
- Does the carrier restrict the speed of its vehicles by mechanical means?
- Does the carrier monitor compliance with speed limits?
- Is the posted speed limit correlated with the conditions of the route?

<u>Routing</u>

- Have designated routes been established for this material?
- Have the following been considered when analyzing the route?

- Road
  - What types and conditions of roads exist from origin to destination?
  - What types of traffic utilize these roads?
  - Do the roads cut through heavily congested areas?
  - What significant hazards may exist (low bridges, weight restrictions, one-lane travel, etc.)?
  - What safety features are present (traffic signals, guardrails, shoulders/breakdown lanes, etc.)?
  - What are the typical speeds along these roads?
  - What is the total trip time and active transit time?
    - Can total transit time be reduced by using team drivers?
    - If not, are adequate and secure layover facilities available along the route?
  - What additional environmental conditions, specifically for times of delivery, may be considered (e.g., ice, fog, low spots)?
- Rail
  - What types of track (track class or rating) exist from origin to destination?
  - What is the maximum speed limit for trains that will carry this product?
  - What safety features are present (double vs. single track, operating signals, automated switch operation, lowered speed in population centers, etc.)?
  - How many times does the container switch trains or carriers?
  - What is the total trip time and active transit time?

4.3.2 Semi-Quantitative Risk Analysis Techniques

Once the specific issues and scope of the analysis are defined, a semi-quantitative risk assessment may be conducted using either risk indexing or a risk ranking matrix. The risk indexing and risk matrix techniques should be built on the information from the earlier analyses. Each level of risk analysis should not be considered a separate effort, but a continued understanding of the transportation issue. Additionally the information gained from these activities can be used to update the qualitative analysis, especially benchmarking comparisons.

4.3.2.1 Risk Index

Risk indices are single numbers or a tabulation of numbers correlated to the magnitude of risk. Some risk indices are relative values with no specific units, which only have meaning within the specific context of the risk index calculation methodology. Risk indices are easy to explain and present, but contain less information than other, more complex measures of risk.

The use of a risk index allows different chemicals, shipments, and segments of a route to be analyzed consistently using a set of predefined categories. These categories are based on the transportation risk concepts presented in Chapter 3. An example of a risk indexing method is shown in Figure 4.3.

In the example shown in Figure 4.3, a score is assigned for each variable (0 to 10) for each chemical product and the scores must be tied to specific definitions under each category. Continuing with the example presented in Figure 4.3, the following hierarchy might be established for the hazard rating:

- Toxic inhalation hazards = 10
- Flammables = 5
- Potential for environmental impacts = 2
- Limited/no impact = 0

The final score or the risk index is the sum of the individual categories (as shown in Figure 4.3) or could be the sum of the individual categories multiplied by additional weighting factors if different categories have more of an influence on the potential risk. Even though the risk index is a simple and helpful tool, the relative weighting of the factors is critically important. Improper weighting can lead to misinformed risk management decisions, a false sense of security, or misallocated resources.

Regardless of the categories selected and scoring system utilized, the end result provides a consistent means of comparing many different transportation activities that quickly identifies:

- Higher-indexed activities
- Possible need for additional analysis
- Comparison of risk mitigation measures

Figure 4.3 Example of Risk Indexing

Product	Frequency of Shipment	Previous Incidents	Quantity in Transit	Hazard Rating	Sensitive Population Along Route	RISK INDEX
Chlorine	5	2	10	10	5	**32**
Gas Odorants	2	1	2	0	10	**15**
Ethylene Oxide	10	5	10	5	7	**37**
Refrigerants	5	10	2	2	5	**24**

4.3.2.2 Risk Matrix

Another useful tool for assisting in semi-quantitative risk decision-making is the risk matrix. A risk matrix is a graphical tool used to identify issues that may require risk reduction as well as those that may require additional detail before an informed decision can be made. Figure 4.4 illustrates a 4-by-4 matrix with risk priorities ranging from I to IV. For example, using Figure 4.4, events that could have very high consequences (consequence category 4) and likely to occur (likelihood category 4) result in a risk priority of I. Since this scenario would suggest a very high level of risk, it would be a candidate for risk reduction.

Before a risk matrix can be used, the consequence and likelihood categories must be developed and guidelines for interpreting the risk results defined. Example consequence and likelihood categories are presented in Tables 4.6 and 4.7 and an example evaluation guideline in Table 4.8. The specific categories, size of the matrix (e.g., X by X), and risk ranking should be company-specific, or agreed to among all supply chain partners involved in the analysis. For example, depending on the chemical properties, mode of transport, and proximity to sensitive populations, it may be appropriate to include multiple impacts in the definition of the consequence categories or to include this higher level of potential impact in the matrix.

Figure 4.4 Risk Matrix

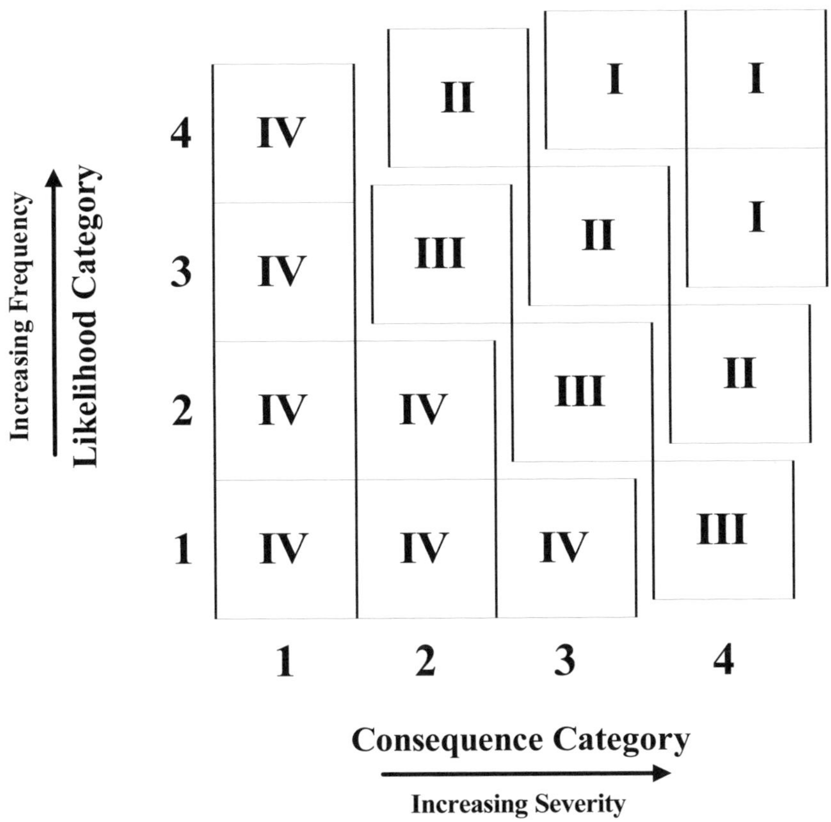

Table 4.6 Example of Qualitative Consequence Categories

Consequence Level	Description
1. Minor	• Safety: No injuries • Environmental: No or small spill response • Property: Damage < $X (units of currency)
2. Low	• Safety: Potential minor injuries • Environmental: Large spill response • Property: $X–$Y (units of currency)
3. High	• Safety: Potential serious injuries • Environmental: Short-term remediation • Property: $Y–$Z (units of currency)
4. Very High	• Safety: Potential fatalities • Environmental: Long-term remediation • Property: Damage > $Z (units of currency)

Table 4.7 Example of Qualitative Likelihood Categories

Likelihood Level	Description
1. Extremely Unlikely (or remote)	Not realistically expected to occur
2. Very Unlikely	Not expected to occur (but credible)
3. Unlikely	Unlikely to occur but has occurred for similar transportation operations
4. Likely	May occur at least once in the lifetime of the transportation operation

Table 4.8 Example of Evaluation Criteria

Number	Description
I	Risk must be mitigated
II	Risk mitigations should be considered
III	Procedures and controls verified
IV	No mitigation required

As shown above, defining a set of evaluation criteria provides a consistent consideration of risk or identification of risks that are candidates for additional risk mitigation. It also enables comparison of risk mitigation options and identification of those options that provide the best level of risk reduction for the capital investment required. Once recommendations have been developed and implemented, all transportation activities are still tracked as part of the overall transportation risk management system.

4.3.3 XYZ Chemical Example—Semi-Quantitative Risk Analysis

Following the qualitative risk analysis of the XYZ Chemical facility located in the Asian market, three chemicals were escalated for a more detailed semi-quantitative risk analysis. These chemicals and modes of transportation are:

- Repackaged chlorine in ton containers
- Bulk and non-bulk pesticide shipments
- Ethylene oxide railcar deliveries

To complete this analysis a corporate risk management representative came to the facility to lead a local team through a semi-quantitative risk analysis methodology. The interdisciplinary team, which had received background information on the risk matrix and the semi-quantitative method that will be used, consists of:

- EHS
- Logistics
- Transportation
- Maintenance
- Regulations

The shippers of chlorine, pesticide, and ethylene oxide were informed that XYZ Chemical was conducting this analysis, and each company made representatives available on an as-needed basis.

As directed by corporate, and consistent with all facilities conducting this type of analysis, the following were defined for the semi-quantitative analysis:

- *Scope of Analysis*—limited to accidental releases of the chemicals and modes escalated from the qualitative analysis
- *Route Detail*—routes will be segmented to identify different representative population densities (designated as High, Medium, and Low) and other receptors along the route(s)
- *Scenarios*—for each chemical/mode of transport, three release sizes will be analyzed for a breach of the transportation container: small hole (i.e., shearing of a valve connection), tear (i.e., puncture of the container), and catastrophic failure (i.e., large breach of the container), and are designated as small, medium, and large releases during the analysis
- *Analysis Guidelines*—for consistency with other facilities, corporate has provided the consequence, likelihood, evaluation criteria, and risk matrix

The 4-by-4 risk matrix for this facilitated analysis is illustrated in Figure 4.5. The consequence, likelihood, and evaluation criteria definitions are detailed in Tables 4.9–4.11, respectively.

Figure 4.5 XYZ Chemical Risk Matrix

Likelihood Category (Increasing Frequency ↑)				
4	*IV*	*II*	*I*	*I*
3	*IV*	*III*	*II*	*I*
2	*IV*	*IV*	*III*	*II*
1	*IV*	*IV*	*IV*	*III*
	1	2	3	4

Consequence Category

→ Increasing Severity

Table 4.9 XYZ Chemical Consequence Categories

Consequence Level	Description
1. Minor	• Safety: No or limited minor injuries • Environmental: No impact • Property: Damage < US$500,000
2. Low	• Safety: Potential multiple minor injuries or limited serious injuries • Environmental: Spill response required • Property: US$500,000 –US$1 million
3. High	• Safety: Potential multiple serious injuries or limited fatalities • Environmental: Remediation for non-sensitive receptor • Property: US$1 million–US$10 million
4. Very High	• Safety: Potential multiple fatalities • Environmental: Remediation for sensitive receptor • Property: Damage > US$10 million

Table 4.10 XYZ Chemical Likelihood Categories

Likelihood Level	Description
1. Extremely Unlikely (or remote)	Not realistically expected to occur
2. Very Unlikely	Not expected to occur (but credible)
3. Unlikely	Unlikely to occur but has occurred for similar transportation operations
4. Likely	May occur at least once in the lifetime of the transportation operation

Table 4.11 XYZ Chemical Evaluation Criteria

Number	Category	Description
I	Action Required—High Priority	Risk mitigation options recommended
II	Action Required	Risk mitigations should be considered
III	Tolerable with Controls	Procedures and controls verified
IV	Tolerable as is	Baseline program element review

Before each chemical/movement was evaluated by the team, the route was reviewed and segmented based on a coarse-level evaluation of the population density. Then, the corporate lead categorized the population as high, medium, or low, based on levels predefined for these worldwide risk analyses. A graphic developed by the team representing the routes is depicted in Figure 4.6.

Figure 4.6 Segmented Routes (Chemicals/Modes of Concern)

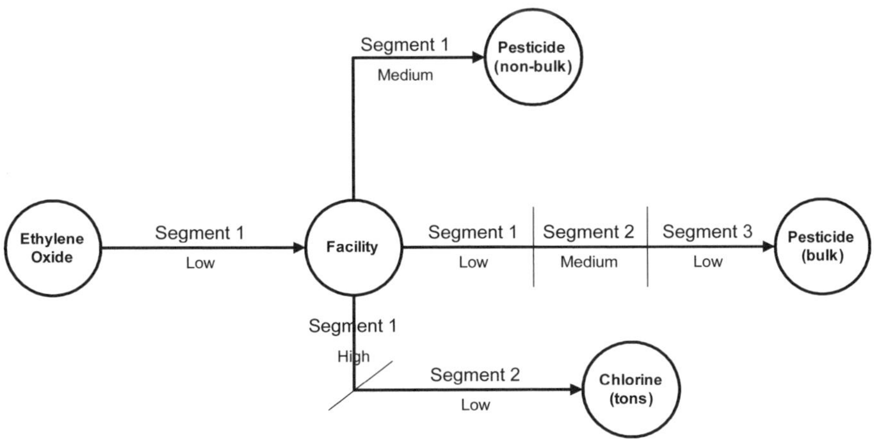

From this figure the following is determined:

- The ethylene oxide route to the facility can be represented by a single "low" population density.
- The pesticide (non-bulk) route from the facility to its destination can be represented by a single "medium" population density.
- Repackaged chlorine ton container shipments from the facility are represented by two segments: Segment 1 "high" population density and Segment 2 "low" population density.
- Pesticide in bulk is represented by two "low" and one "medium" population density segments.

Using the consequence and likelihood categories, risk matrix, and risk evaluation criteria, the team reviewed three release scenarios (small, medium, and large) for the segments identified for each of the chemical movements. The result of the semi-quantitative risk estimation for this facility's hazardous material transportation operation is detailed in Table 4.12. From this results table, the following are determined:

- Ethylene oxide railcar receipts and the pesticide (non-bulk) shipments were identified as lower risk and no additional risk mitigation measures were recommended by the team. Risk ranking of III and IV, respectively.
- Chlorine ton container shipments through Segment 2 (low population density) were identified as lower risk. Risk ranking of III–IV.
- Chlorine ton container shipments through Segment 1 (high population density) were identified as a higher risk, with action required. Risk ranking of II.
- Pesticide (bulk) shipments were identified as a higher risk, with action required for the low population density segments (risk ranking of II –III) and priority action for the medium population density segment (risk ranking of I–II).

Table 4.12 Facility Risk Analysis Results

Scenario ID	Chemical	Release Size	Population Density	Consequence Level	Likelihood Level	Risk Ranking
EO-s-low	Ethylene Oxide	Small (s)	Low	2	3	III
EO-m-low		Medium (m)		2	3	III
EO-l-low		Large (l)		3	2	III
P-nb-s-med	Pesticide (non-bulk)	S	Medium	1	3	IV
P-nb-m-med		M		1	2	IV
P-nb-l-med		L		3	1	IV
Cl-s-high	Chlorine (tons)	s	High	2	4	II
Cl-m-high		m		3	3	II
Cl-l-high		l		4	2	II
Cl-s-low	Chlorine (tons)	s	Low	1	4	IV
Cl-m-low		m		2	3	III
Cl-l-low		l		3	2	III
P-b-s-med	Pesticide (bulk)	s	Medium	3	4	I
P-b-m-med		m		4	3	I
P-b-l-med		l		4	2	II
P-b-s-low	Pesticide (bulk)	s	Low	2	4	II
P-b-m-low		m		3	3	II
P-b-l-low		l		3	2	III

The results of this risk analysis indicate that ethylene oxide, non-bulk shipments of pesticide, and the segment of the chlorine (ton containers) routed through the low population density can be eliminated from further review. The team also decided that no additional risk mitigation actions were necessary for these transportation activities (and were not required based on the risk ranking). These chemicals and modes of transport will, however, continue to be reviewed and monitored as part of the global risk management program.

For the segment of the chlorine (ton containers) routed through the medium population density and the entire bulk pesticide transportation route, the team developed recommendations for consideration. These recommendations included the following and are evaluated in Chapter 7:

- Chlorine:
  1. Consider rerouting truck deliveries to bypass the population center as this segment of the route is not necessary for reaching the final destination.
  2. Consider changing the shipment from ton containers to smaller cylinders.
- Pesticide (bulk):
  1. Consider rerouting tank truck deliveries to eliminate passing near higher population densities along the route.
  2. Consider decreasing the tank truck shipment volume and increasing deliveries.

Even though higher-level risks were identified for two of the chemical movements, this level of analysis lends itself to recommending and evaluating risk reduction strategies, as warranted. In this continuing example, the bulk pesticide is escalated for quantitative analysis (Chapter 5) and the chlorine containers risk reduction options are the focus of Chapter 7.

REFERENCES

Center for Chemical Process Safety, *Guidelines for Chemical Process Quantitative Risk Analysis, Second Edition*, American Institute of Chemical Engineers, New York, 2000

Center for Chemical Process Safety, *Guidelines for Chemical Transportation Risk Analysis,* American Institute of Chemical Engineers, New York, 1995

Lees, F.P., *Loss Prevention in the Process Industries, Second Edition*, Butterworth Heinemann, London, 1996

5
QUANTITATIVE RISK ANALYSIS

This chapter provides an overview of quantitative risk analysis (QRA) techniques for evaluating hazardous materials (or dangerous goods) transportation issues, including data requirements, analysis techniques, and the generation and interpretation of quantitative risk results. The intent of this chapter is not to be the ultimate technical resource for transportation data, consequence modeling, frequency analysis, or calculation of quantitative risk results, but rather to serve as a resource for the transportation professional participating or supporting a QRA. *Guidelines for Chemical Process Quantitative Risk Analysis* (CCPS, 2000) is the technical resource on applying a QRA, and *Guidelines for Chemical Transportation Risk Analysis* (CCPS, 1995) is the technical resource for adapting these fixed-facility techniques to address the unique issues of hazardous material transportation.

This chapter is not meant to suggest that rigorous analysis techniques should be employed in all cases, and is only intended for advanced risk analysis applications. The goal of this practical review of transportation QRA is to help the reader better understand:

- Basic QRA concepts and terminology
- QRA calculation techniques
- Data requirements
- Data limitations
- Results and presentation formats
- Common pitfalls

As stressed throughout this book, and is common industry practice, the escalation to a QRA should be used sparingly and only for the toughest risk management problems. Simpler qualitative and semi-quantitative techniques presented in Chapter 4 should be exhausted, with the additional detail provided through a QRA sought only when required to fully understand and compare the level of risk or to evaluate and justify risk mitigation options. Although a QRA can contribute valuable information to the decision process, this technique (as compared to qualitative and semi-quantitative) is the most resource- and data-intensive. Therefore, when pursuing a QRA, care must be taken to fully understand the scope of the analysis and ensure that quality data can be obtained and applied correctly. Otherwise, a costly analysis, in the end, may not provide the clarity or confidence that was envisioned.

Section 5.1 provides an overview of QRA, including considerations for use, study design, and protocol. Data requirements for frequency analysis, consequence analysis, and other select QRA parameters are presented in Section 5.2. The interpretation and presentation of QRA results are illustrated in Section 5.3. To bring these concepts together, the XYZ Chemical example uses QRA to evaluate the issues escalated from the qualitative and semi-quantitative analyses covered in Chapter 4. Further, Chapters 6 and 7 of *Guidelines for Chemical Transportation Risk Analysis* (CCPS, 1995) (see Appendix A on enclosed CD) contain additional examples and case studies that illustrate the application of QRA.

5.1 OVERVIEW

Similar to the process defined for semi-quantitative risk analysis in Chapter 4, quantitative risk is a function of consequence and likelihood. Decisions are made regarding the resulting risk level and the need for additional risk mitigation based on a comparison of the results to a set of criteria. The QRA process is illustrated in Figure 5.1. The differences between a QRA and the other simpler techniques are:

- Scope is based on the issues escalated through the use of simpler techniques.
- Data requirements to complete the analysis can be significantly greater.
- Supplemental and better quality data are critical inputs to a more detailed analysis.
- Frequency and consequence are both analyzed in more detail than during a semi-quantitative analysis.

Before a QRA is pursued, it is important to ensure that the simpler risk analysis techniques, and their analysis options, have been conducted to the fullest extent. A QRA will provide benefits to the decision process if:

- The risk estimate requires additional refinement of consequence and/or likelihood.
- Simpler risk analysis approaches have left too many options.
- Decisions cannot be made or there are unanswered questions/issues.
- Valuable insight into the "risk drivers" is needed to identify the selection of appropriate risk mitigation options.
- A cost-benefit analysis of the relative difference between options needs to be evaluated.
- Corporate management or a regulatory agency demands a defensible analysis and a clear rationale for estimating risk.

Figure 5.1 Quantitative Risk Analysis Process

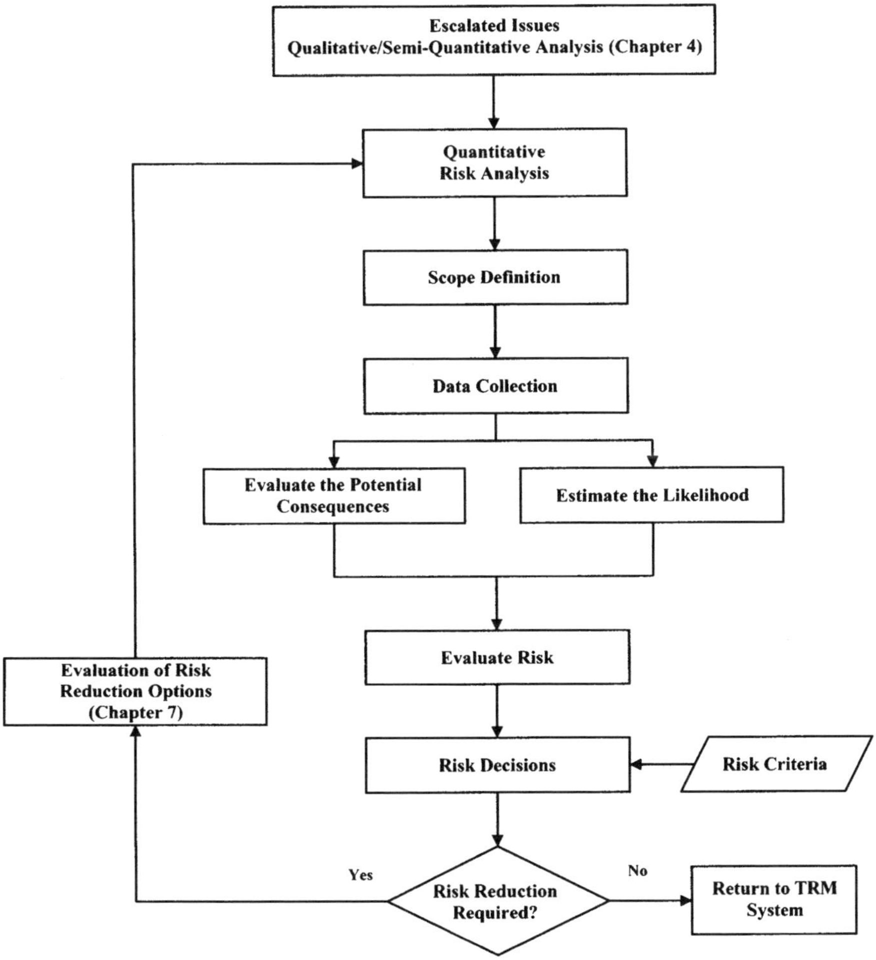

5.1.1 QRA Study Design

Transportation risk assessments can range from a single chemical movement along a specific route segment to a review of an entire transportation network. Therefore, the scope of a QRA should be clearly defined before starting down the risk analysis path. While a QRA can accommodate the most complex transportation system, it is best used on focused issues that have been escalated from the simpler risk analysis techniques while keeping in mind how the results will be used in the overall decision process. Focusing the QRA will ensure an efficient use of resources (personnel and funds) and, if required, the study can be expanded. Parameters that may be included in the QRA design may include, but are not limited to:

- Comparison of modes of transportation

- Quantity, or package sizes, in transit
- Chemical-specific hazards
- Route characteristics
- Transportation package design

The most appropriate depth of study for a given situation is determined by considering not only the study objectives, but also the data requirements and the availability of all necessary information. If only limited or generic data are available, a high degree of quantitative differentiation between two options may not be achievable, and a very detailed depth of study may not be warranted. Similarly, if selected data sets are considered to be very uncertain, it may not be worthwhile to be as precise as is possible in other parts of the analysis.

QRAs are also conducted differently than the simpler risk analysis methods. Simpler risk analysis techniques are typically conducted in a team environment, with representative stakeholders participating and contributing to the entire evaluation. With a QRA, as the complexity increases, the team member's role changes from conducting the analysis to supporting the analysis through guidance, data collection, and validation of the results. In practice, QRAs need to be conducted by risk professionals with experience in the methodology, consequence analysis, frequency analysis, and interpretation and presentation of risk results. Depending on the company, these specialized people may be internal or external to the organization and the supply chain partners.

Compared to fixed-facility QRAs, transportation QRAs can be even more variable in terms of resource requirements because a study can cover a single chemical movement or many such movements within a single analysis. In addition, each time an alternate transportation option is included for evaluation, it is comparable to evaluating a whole new processing unit at a fixed facility.

5.1.2 QRA Protocol

Once the study scope is finalized, a protocol should be developed and followed outlining the steps for completing the QRA. An example protocol for conducting a QRA may include the following steps:

- *Step 1—Selection of Scenarios/Issues for Evaluation*: This step should be in alignment with the scope of the QRA and includes defining the modes of transport, hazardous material movements, package designs, and routes or route segments that will be evaluated or compared.
- *Step 2—Data Collection and Evaluation*: The second step in conducting an analysis of this type is the identification and collection of data necessary for the detailed risk assessment. A variety of methods can be employed to gather the required information including a review of previous studies, an extensive search and review of literature in the public domain in the study field, a review of confidential data, discussions with carriers and other stakeholders, and other sources.

- *Step 3—Selection of Consequence Measures*: Transportation risks arise from the release of hazardous materials as a result of accident- and/or non-accident-initiated events. The types of impacts that may be evaluated in a QRA include impacts to people, property, the environment, disruption of supply chain activities, and other financial concerns.
- *Step 4—Perform QRA*: This step involves using the scenarios defined in Step 1 and the data in Step 2 to conduct the frequency analysis, consequence analysis for the defined impacts of Step 3, and generation of the quantitative risk results.
- *Step 5—Presentation of Results*: The results of a QRA are used for making informed decisions. Therefore, the selection of risk measures and presentation formats should meet the objectives of the study and be tailored for the intended audience. The presentation may include the use of risk criteria, comparison to other options, and cost-benefit analysis. Often, and as presented in this chapter, quantitative risk results are presented in the form of a risk curve.

Two additional steps that may be included as part of a QRA include sensitivity analysis and the evaluation of risk reduction options. Sensitivity analyses are often used to evaluate the influence of various data components and assumptions of the QRA. One should always remember that the risk analyst is dealing with risk estimates, and, in order to use these estimates properly, it is essential that the potential extent of uncertainty or key assumptions that are a major influence on the risk results be known and understood.

If it is determined that the estimated level of risk should be reduced, the evaluation of risk reduction options is performed. This step is the focus of Chapter 7, and for each option selected, Steps 1–5 from above need to be completed so that the results for the options can be compared on a consistent basis.

5.2 QRA DATA SOURCES

The availability and proper use of data are critical to a successful QRA. QRA data sources for the different modes of transportation are detailed in *Guidelines for Chemical Transportation Risk Analysis* (CCPS, 1995). The focus of this chapter is to provide an overview of the different types of data required to complete a QRA and to point out the potential pitfalls that should be considered when using transportation data.

Examples of the types of data that may be required (depending on mode) to generate the results of a QRA are illustrated in Figure 5.2. In general, there are five main categories of data that are required to generate QRA results or the risk profile of the operation being analyzed. These categories and their specifics will vary with mode of transportation and may include:

- *Commodity Data*—information about the hazardous material(s) that is being shipped such as physical properties of the chemical, packaging types (including quantity in transit and storage conditions), annual shipments, time of day the shipment moves across the route or segments of the route, and breakout of loaded versus empty miles

- *Route Data*—information regarding the route and includes conditions of the infrastructure (e.g., road and rail condition and speed restrictions), specific segments of the route identified for analysis (including meteorological conditions), and the location of sensitive receptors along the route (e.g., urban and rural population densities, environmental crossings)
- *Demographic Data*—the description of the population along the route, which can include the development and use of representative urban and rural population densities along the route or more precise estimates (e.g., U.S. census data)
- *Frequency Data*—the data required to generate accident and non-accident rates, the probability of a release following an accident, the range of release sizes to be considered, and the probabilities of various outcomes of release (i.e., toxic, flammable, explosive)
- *Consequence Data*—the information on the chemicals, mode of transportation, and the potential release sizes, consequence models is used to determine the extent of the release; combining the release extent with potential receptors surrounding the route results in the estimated number of impacts

Figure 5.2 QRA Data Requirements

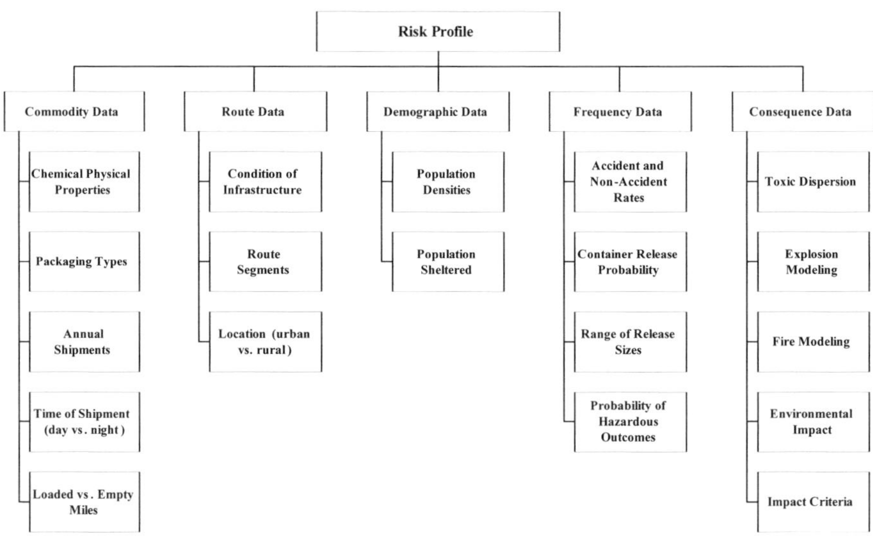

Data quality and uncertainty are present at all levels of risk analysis (qualitative, semi-quantitative, and full quantitative). Increasing the level of analysis detail may not result in a more accurate estimate of risk and, in fact, going to a deeper level of analysis may actually increase the uncertainty. When conducting or participating in a QRA, the following questions should be asked about the data:

- *Are data available for the defined scenarios*? If data are not available, or if reasonable and verifiable assumptions cannot be made that apply to the specific transportation operation, then the scenarios may need to be modified. Simple scenarios—meaning high level or broadly representative ones—are usually better for transportation risk analysis. For instance, a scenario could be defined as a large release of a flammable from a rail car in a populated area as opposed to defining the scenario in this way: a flammable rail car derails, rolls down an embankment, crashes into a building at the bottom of the hill, and results in a large release in a populated area. The first scenario, with broader definition, covers the second scenario, as well as many other detailed ones that may be considered.
- *Are data suitable for the analysis*? While there may be numerous sources of published data, the data may not be suitable for use in a risk assessment because the information is not accurate, representative, or even appropriate for the specific analysis.
- *What is the level of confidence in the data*? Data should be reviewed with other stakeholders and the carriers.
- *How will the results be interpreted*? The study design should have the scope, objectives/questions to be answered, data required, methods of analysis, and decision criteria all defined prior to starting a QRA. One should not go too far into a study before determining how the data will be analyzed and the results will be interpreted..
- *Are the data accurate and detailed enough to justify predicting the desired consequence levels (e.g., fatalities, injuries, evacuations, environmental, economic)?* If there is not a very high level of confidence in the data, the analysis may need to adjust the consequence measure (i.e., from fatalities to number of people who might be evacuated), change the scenarios, or go back to a lower level of analysis.

5.2.1 Data for Frequency Analysis

An important component in a transportation QRA is a frequency analysis. This includes estimating the frequency of accidents, the chance of release (if there is an accident), the range of possible release sizes, and the probability of various outcomes if a hazardous material is released during transit. The general principles that should be considered when calculating the accident rate may include the following:

- *Use the most applicable, statistically sound data for your situation.* For example, if a specific truck movement is being analyzed, it would be best if the accident rates of the companies contracted for that shipment are used. Where available, use accident rates of the specific carriers along the specific routes under evaluation, but only if the carriers have enough miles so that sound and justifiable accident rates can be developed.

- *If only generic accident data are available, pick them carefully and ensure that they apply reasonably well to the situation being analyzed.* For example, if hazardous material tank truck movements are being reviewed, do not apply generic truck accident rates developed for all freight. In most cases, this will result in significant overestimation of accident likelihood. Hazardous goods are generally transported with more care and attention. Consequently, more applicable accident rates over the same route may be lower.
- *Align level of detail with the quality of the data.* For example, segment-specific accident rates should not be applied to rail or highways unless good data exist for both numerators (number of accidents) and denominators (number of miles). Also, ensure that there have been a sufficient number of miles traveled across a segment so sound and justifiable accident rates can be developed.
- *Be cautious about using rates where the numerator and denominator are derived from different sources.* In these cases the number of accidents, number of miles traveled, and time frame may not align.
- *Ensure that the data are representative of the transportation operation.* Be cautious about using data from certain sources (i.e., a specific county or consolidated data for a specific mode) that might not be representative of the specific transportation network being studied. For example, such data may aggregate the information from all types infrastructure (road conditions, speed restrictions, etc.), and your situation might only involve movement along a new and uncongested segment of road.

The chance of an incident is generally a function of the distance traveled. Thus, the frequency of an accident is often expressed as an accident rate per mile. Contributions from non-accident-initiated events are typically expressed on a frequency-per-hour or per-year basis. Thus, the duration of the hazardous materials movement is a key parameter. Figure 5.3 illustrates the basic calculation sequence for one trip or movement. If multiple trips are made, the total risk is equal to the number of trips times the risk per trip. The basic calculation sequence will have minor variations for each mode of transport and can be broken down into greater detail as needed. Increased detail might include different accident rates and lengths for each segment of a route or might explicitly address the accident rates and release probabilities for different accident causes. Inputs to the analysis that may be altered or may influence the calculation include:

- Accident data
- Mode of transportation
- Geographic location
- Transport route considerations, such as:
  - Speed limits, congestion, actual speed, traffic, infrastructure condition
  - Carrier specific data
  - Container specifications

Figure 5.3 Calculating the Chance of Release

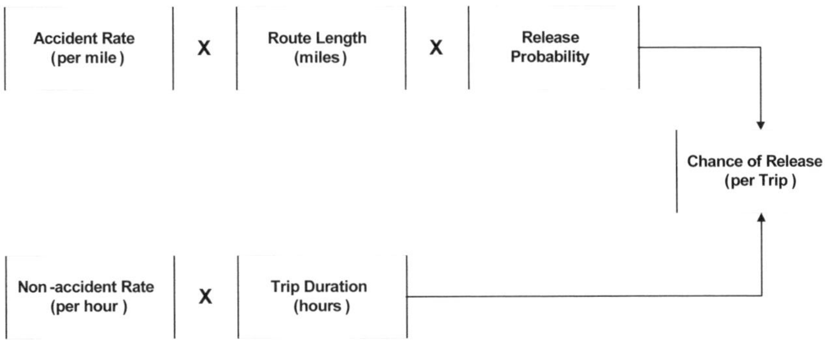

In addition to determining the chance of release as a result of an accident or non-accident (the conditional probability of release), it is important that the analysis considers some measure of the severity of the release. Very small releases may result in only minor impacts, while larger, less frequent releases could pose a considerable hazard. This effect is incorporated into the analysis using a spill size distribution. This distribution is the range of possible release sizes (e.g., small-minor, large-severe) and the probability of each given an incident has occurred. Additionally, a spill size distribution should be specific to the modes of transport, type/severity of the accident, and container design specifications. Once the chance of release in Figure 5.3 has been developed, the likelihood of release for each hole size can be calculated by multiplying the chance of release by the corresponding spill size distribution probability. In a typical QRA, not every possibility is considered; only a representative range of small to large release sizes is evaluated.

Event tree analysis can be used to complete the frequency calculations. An event tree is a graphical logic model that identifies and quantifies possible outcomes following an initiating event. For transportation, these could be the outcomes where an accident has occurred or the outcomes following a non-accidental release. To complete an event tree, the initial release frequency (chance of release times the release size probability) goes on the initiating branch. Next, probabilities are assigned to the remaining branches of the event tree, which are associated with the probability of the different outcomes. The final result is a unique scenario frequency that will be paired with each potential consequence. Figure 5.4 shows an example event tree. The probabilities populated in a event tree are dependent on the chemical released and the surrounding environment along the route.

The event tree illustrated in Figure 5.4 was developed specifically for the post-incident outcomes, initiated with the release and ending with the final consequences. Event trees can be developed to characterize the pre-incident (the events leading up to an accident or non-accident release), the post-incident outcomes, or the entire scenario under evaluation.

Figure 5.4 Example Event Tree

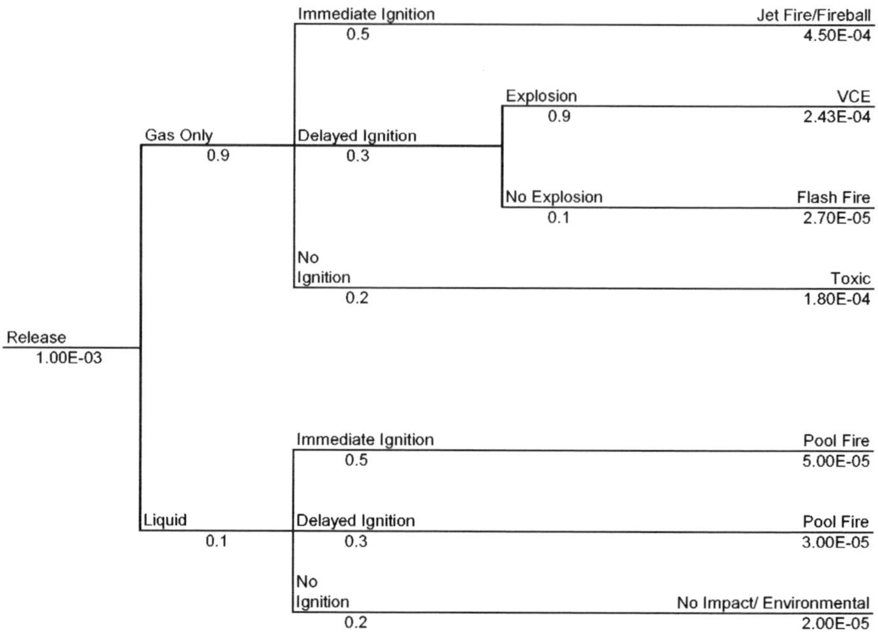

5.2.2 Data for Consequence Analysis

In the terminology of QRA, consequence is a measure of the expected outcome of an event and is measured or expressed as "hazard distances" or "hazard zones." The impact analysis addresses the potential effects of the hazardous consequences to people, property, and the environment. A consequence analysis is normally carried out using mathematical models and computer software addressing the physical and chemical phenomena. Impact analysis takes into account the presence of people, property, and sensitive environments that can be adversely be affected by a chemical release.

Before using a consequence assessment model, it is necessary to postulate events that follow the release of a hazardous chemical. In general, a hazardous material release may exhibit one or more of the following types of hazards:

- Flammable exposure (thermal radiation, impingement)
- Explosions (blast overpressure, fragments)
- Toxic exposure hazards
- Environmental contamination

When modeling the hazard zones or distances for different types of hazards, it is very important to use a consistent set of criteria. These criteria, often called "endpoint" criteria, refer to the following:

- *Toxic Chemical Exposure*—the endpoint criterion is defined as a limiting concentration, often expressed in ppm or mg/m^3, over an exposure duration. Typical exposure durations may range from a few minutes to a period of several hours.
- *Vapor Cloud Explosion*—the endpoint criterion is defined as a limiting blast overpressure, often expressed in psi or bar, which could result in substantial damage to buildings, structures, or people.
- *Flammability Hazards*—several endpoint criteria are used. For flammable liquids that form large pool fires, a steady heat load or thermal radiation criterion often expressed in BTU/hr-ft^2 or kW/m^2 is used. For determining hazards from short-duration BLEVEs and fireballs, an integrated dose criterion, which represents the area of a time-dependent heat flux, is used. For determining flash fire hazards, a flammable vapor has to be diluted to within its flammable range, with the concentration often usually expressed in ppm, mg/m^3, or as a volume percent.

The hazard distances are directly influenced by the endpoint criteria used in the consequence analysis, with lower criteria leading to larger affected areas. Therefore, it is important to establish a level of consistency in using these criteria. Chapter 2 of the *Guidelines for Chemical Process Quantitative Risk Analysis, Second Edition* (CCPS, 2000) should be referenced for selecting appropriate and consistent toxic, explosive, and flammable endpoint criteria.

5.2.2.1 Factors Affecting Fire and Flammability Hazards

Upon ignition, spilled flammable material may result in a pool fire, jet fire, or flash fire depending upon the material properties and process conditions. The magnitude of the resulting consequences will depend upon the release quantity and the chemical properties. The hazardous impact area in this type of situation should also consider thermal radiation effects. The actual radiation received by an object or person depends upon the distance from the flame surface, humidity, and other atmospheric conditions. The radiation impact at a given location also depends upon the orientation of the person or structure to the flame, with sheltering potentially reducing the magnitude of the hazard.

For pool fires, if the spill is confined, the area of the pool will determine the maximum flame area/size. If the spill is unconfined, the spill dimensions will depend on the liquid volume and the burning rate of the liquid.

Jet fires typically occur when a pressurized material is released and ignited at the source. Instantaneous releases of a pressurized material coupled with immediate ignition may lead to the formation of a fireball. Delayed ignition of a pressurized flammable may lead to a flash fire or explosion, both of which are discussed in the next section.

5.2.2.2 Factors Affecting Flash Fire/Explosion Hazards

The release of a flammable material may lead to a vapor cloud explosion (VCE) upon ignition. A VCE results from the rapid combustion of a fuel/air cloud with the flame speed approaching sonic velocity, thereby producing a blast wave. The explosion potential of a flammable hydrocarbon will depend on the combustion energy of the material and the energy of the ignition source. In addition, the fraction of the combustion energy converted to the explosive energy will depend on the nature of the material that has been ignited.

Turbulence is required for the acceleration of flame front to speeds required to produce the blast overpressure associated with a VCE. In the absence of turbulence, a flash fire will occur without any appreciable overpressure, with the hazard limited to the thermal radiation impacts associated with the burning of the cloud from the ignition point back to the release source, or within the flammable range of the cloud. Flame turbulence is typically formed by the interaction between the flame front and obstacles. The blast effects produced by VCEs vary greatly and are primarily dependent on flame speed; therefore, areas of confinement and congestion near the release point can influence the likelihood of a VCE. Additionally the reactivity of the material is an important consideration; highly reactive materials such as ethylene oxide are much more likely to lead to a VCE than lower reactive materials such as methane.

5.2.2.3 Factors Affecting Toxic Hazards

The release of a toxic chemical might produce little, if any, property damage, but could result in significant impacts to a population surrounding the release point. Some chemicals pose acute inhalation hazards and can result in damage to the respiratory system or other critical functions. Other chemicals are carcinogenic in nature, and the impact caused by a single exposure to these materials can result in an increased potential for developing cancer later in life. In general, toxic vapor dispersion hazard zones may be characterized by the chemical properties of the material released as well as by the following parameters:

- Release quantity
- Duration of release
- Source elevation
- Surrounding terrain
- Prevailing atmospheric conditions
- Limiting concentration (downwind concentration profile)
- Sensitivity of the surrounding ecosystem

Each of these parameters is discussed below with special emphasis on its influence on estimating downwind dispersion distances. It is not practical to consider all possible combinations of these factors, because an incident could occur at any point along the route. Therefore, representative parameters, or sets of parameters, are typically developed. Specific parameters can be used if an exact location along a route is being analyzed, and the required data is available.

- *Release Quantity* or release rate refers to the quantity of (or the rate at which) a hazardous chemical is released from it container in an accident. The quantity (or rate) is the single most important parameter in determining dispersion hazard distances. In general, larger quantities lead to larger dispersion distances. However, the resulting dispersion distance may not increase linearly with quantity or release rate. For instance, an increase in release quantity by a factor of 100 (for example, from 100 gallons to 10,000 gallons) may only increase the dispersion hazard distance by a factor of 10 or 20. For gaseous and high-vapor-pressure liquid releases, the vapor release rate will be the same as the discharge rate. However, for non-flashing liquids, the vapor release rate is governed by the evaporation rate of the liquid and will always be less than the liquid release rate.
- *Duration of Release* is dependent on the release mode and system inventory of material. For example, a safety valve release may not last for more than a few minutes whereas a tank puncture may continue to discharge for several hours. For liquids forming evaporating pools, the duration of the vapor release is dependent on the evaporation rate. Simple dispersion models use one of two extreme cases (i.e., instantaneous release or continuous release). In the case of instantaneous release, the duration of release is very short (e.g., pressurized tank rupture) and the total quantity of the chemical released during the accident contributes most significantly to the dispersion hazard. In the case of a continuous release, the release duration lasts a relatively long time and the release rate is the most significant parameter.
- *Source Elevation* is attributed to the physical height of the release. A higher source elevation will increase the distance to cloud touchdown at ground level. This is particularly important for buoyant and neutrally buoyant clouds, since this may minimize ground-level impacts.
- *Surrounding Terrain* greatly affects the dispersion process. For example, rough terrain involving trees, shrubs, buildings, and structures usually enhance air entrainment. This leads to shorter dispersion distances than a release that occurs in open or relatively flat terrain.
- *Prevailing Atmospheric Conditions* include a representative wind speed and atmospheric stability. The neutral or unstable (typical daytime) weather conditions generally lead to shorter dispersion distances than stable (nighttime) weather. For neutrally buoyant releases, increased wind speed reduces the dispersion distance. For heavy gases, dispersion distance increases with wind speed. Since the weather conditions at the time of an accident cannot be controlled or predicted in advance, it is important to evaluate the release scenarios for a range of typical weather conditions.
- *Limiting Concentration* affects the dispersion distance inversely. Lower limiting concentrations lead to larger dispersion distances. As with the source release rate, the effect is not linear. In other words, a factor of 100

reduction in the limiting concentration may result in an increase in the dispersion distance by only a factor of 10 to 20. When selecting endpoints, or levels of concern, it is important to make a distinction between concentrations where there will be some observable effect, and concentrations at which one can expect serious ill effects.

5.2.3 Other QRA Data Categories

In addition to the data elements that go into the frequency and consequence analyses detailed above, there are two other categories of data that will be directly discussed. These include the selection and use of meteorological data in the consequence analysis and the population data for the evaluation of impacts. The reader may refer to Chapter 2 of the *Guidelines for Chemical Transportation Risk Analysis* (CCPS, 1995) for additional information on these data sources as well as any others not explicitly discussed in this chapter.

5.2.3.1 Meteorological Data

The outcome of a hazardous material release is highly dependent on the local meteorological conditions at the time of release. It is a necessary and important part of any risk analysis to select appropriate weather conditions for performing the dispersion consequence modeling. While rigorous observance of meteorological conditions adds unnecessary complications without necessarily providing more accurate results, the selection of unrepresentative meteorological data may lead to gross underestimation or overestimation of the consequences, and hence, an incorrect evaluation of risk.

In evaluating the consequences of a release, there are two meteorological parameters of significance: the prevailing wind speed and the atmospheric stability. A lower wind speed usually leads to slower dilution and larger hazard areas. The stability refers to the ability of the atmosphere to dilute a released chemical. These are classified as Pasquill stability categories ranging from A (highly unstable) to F (highly stable). The late afternoon hours are typically categorized as A or B, whereas the calm hours of the night or early morning are usually given an E or F. The stability determines the "slowness" of dilution, or the speed at which the concentration in the cloud decreases. As such, F stability usually leads to very large dispersion distances. The difference in the dispersion distance for F stability and A stability, with everything else being equal, could easily be an order of magnitude. It is common practice in transport risk analysis to combine the frequencies of A through D stabilities and use the dispersion calculations of D stability (typical or neutral conditions) and combine the E and F stability frequencies and use F stability in calculations. This procedure usually leads to a conservative estimate of risk with more discrete wind speed and stability class combinations factored in to the analysis.

5.2.3.2 Population Data

Population data are often incomplete and not always found locally at a level of detail to accurately estimate the risk along a transportation route. If this is the case, there are several sources that may provide valuable information. In the United States, data can be obtained from the Census Bureau in varying degrees of detail. This type of information is also available internationally. Data are required on the presence and location of people surrounding the transportation route in order to calculate the possible number of impacts (e.g., potential fatality, injury, or evacuation levels).

In practice, representative population densities are applied to the various sections of the route being evaluated. These densities should be consistent with the population that surrounds the route and can be developed for:

- Major cities
- Urban areas
- Suburban areas
- Rural areas

5.3 PRESENTATION OF QUANTITATIVE RESULTS

No one measure is complete or adequate for conveying all the available information about the risk of a transportation activity. Therefore, the most appropriate measure or measures will depend on the:

- Nature of the issue being examined
- Audience for the results
- Format of the previous results that may be used for comparative purposes
- Information and resources available for the study
- Availability of risk criteria

The relative application of the results includes a ranked comparison of various risk reduction schemes, which is the topic of Chapter 7. In general, fewer options and less potential for risk reduction are available in transportation as compared to what can be achieved at a fixed chemical facility. Given the amount of time required to make certain changes, risk reduction options should be addressed explicitly in both the risk benefits and the cost.

5.3.1 QRA Results

There are three basic types of risk measures that may be developed for a quantitative risk analysis:

- *Risk Indices*—single numbers or scores
- *Individual Risk*—refers to the risk level at any given location from causes originating along the route and can be broadly defined as the frequency at which an individual or groups of individuals may be expected to sustain a given level of harm (e.g., fatality, serious injury) from the realization of specified hazards

- *Societal Risk*—the relationship between likelihood of an event and the number of people affected

Table 5.1 summarizes a number of risk measures defined in the *Guidelines for Chemical Process Quantitative Risk Analysis, Second Edition* (CCPS, 1989) and illustrates the advantages and disadvantages when used for transportation risk analyses. Since measures such as the fatal accident rate (FAR) are geared toward estimating risk to employees at fixed facilities, this measure has not been included as it does not generally apply to transportation risk analysis.

Table 5.1 Risk Measures for Transportation QRA

Measure	Advantage	Disadvantage
INDEX: Average Rate of Death	Provides indication of total risk per year	Does not differentiate between high likelihood/low consequence and low likelihood/high consequence events; may not be absolute criteria for tolerating or rejecting risk
INDIVIDUAL RISK: Individual Risk Contour	Illustrated geographical distribution of individual risk; gives indication of number of people at various risk levels	Very time-consuming to produce along an entire route
Maximum Individual Risk	Illustrates highest risk to any one individual from transportation activity, generally directly along route	Does not consider how many people are at this risk level or how quickly the risk level changes with distance from the route
Average Individual Risk—exposed population	Given use of some simplifying asumptions regarding population density, this meaure is often more applicable to transportation risk than for fixed facilities; can be compared to fixed facility and background risks and to risk critieria	Does not indicate number of individuals involved in any one accident; does not indicate maximum risk level; may be misleading when comparing risks averaged over a few people versus those averaged over many people due to large hazard distances
Average Individual Risk—total population	Allows risk to a constant population to be compared for several movements; allows for the extent of consequence to figure more heavily in risk measure	May be significantly misleading with regard to the maximum risk level if this risk is limited to a small portion of the defined community
SOCIETAL RISK: F-N Curve	Displays information about high likelihood/low consequence and low likelihood/high consequence events	Illustrations of total risk along a route or of maximum consequence event may be difficult to interpret

5.3.1.1 Individual Risk

In evaluating measures of individual risk, it is necessary to specify at the outset whether the calculated frequencies relate to those most at risk from a given activity (for example, as a result of their location, habits, or time periods for which they remain vulnerable) or whether they relate to the average value, which is representative of all potentially affected individuals. The former individual risk measure concerns maximum risk levels for an individual or a group such as populations located along the transportation route.

In contrast, average individual risk is more commonly used to provide some broader perspective on the analytical results. This may be achieved by comparing the estimated average risk levels against those arising from a range of common activities or occupations. Clearly, the use of such averages is only appropriate where the risk is relatively uniform over the relevant population. These measures can otherwise be highly misleading. For example, if few individuals are exposed to high risk levels, this would be missed by averaging their risk with a large number of additional individuals at low risk.

Risk along a transportation route is generally represented as linear rather than as point sources, a function of its moving from origin to destination. As illustrated in Figure 5.5, a risk contour or risk transect can be used to illustrate the individual risk as a function of the distance from the route. A risk transect is specific for displaying individual risk for transportation risk results, as compared to a risk contour surrounding a fixed facility. A transect is developed by summing the likelihood of impact for each scenario at each location. Individual risk contours can be overlaid onto a plot plan or area map to visualize the level of risk. While risk contours do not show the number of individuals exposed, they are very useful in geographically indicating the most exposed population.

Figure 5.5 Example Risk Contour

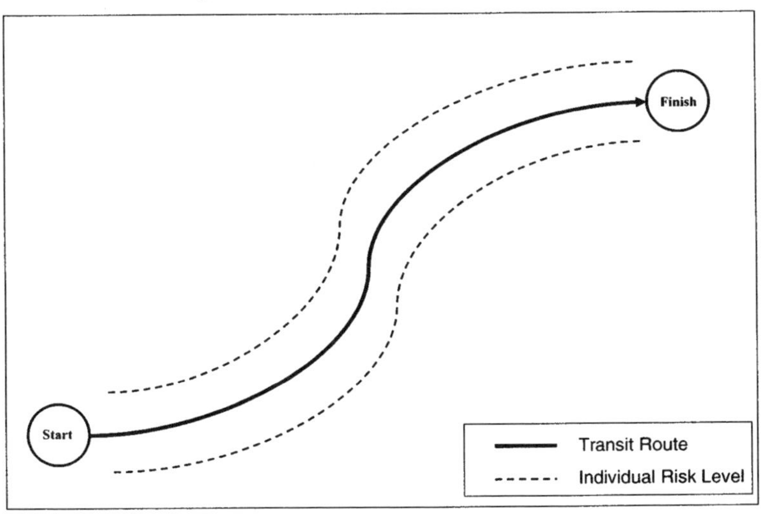

Since it is often difficult to present individual risk as an overlay of the transit route, due to the route length, it is possible to report the risk transect using individual risk as a function of the distance from the route. This type of comparison is shown graphically in Figure 5.6.

Figure 5.6 Example Transportation Risk Transect

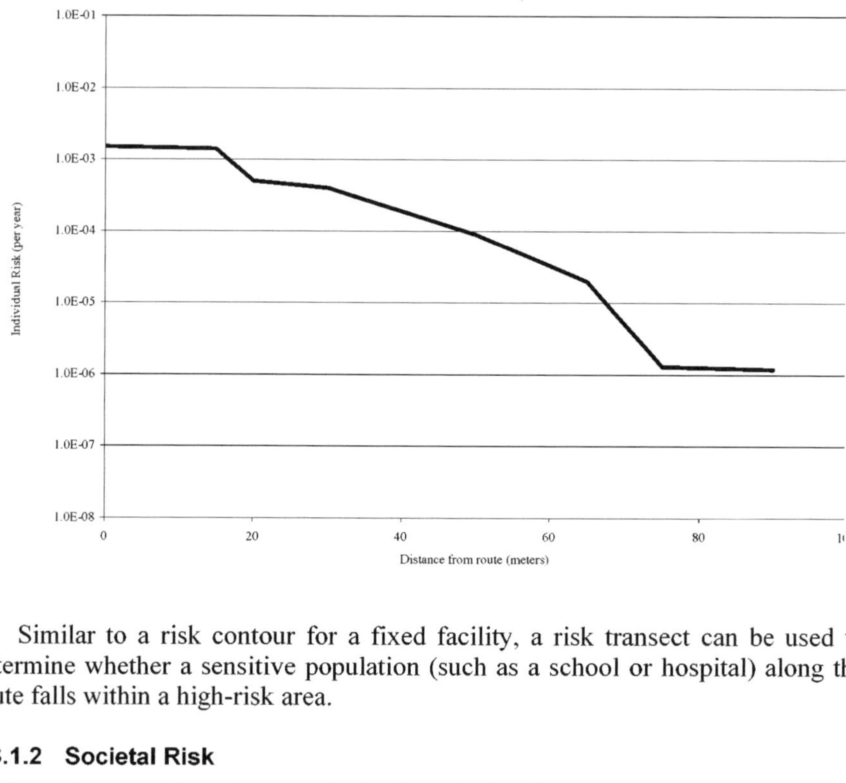

Similar to a risk contour for a fixed facility, a risk transect can be used to determine whether a sensitive population (such as a school or hospital) along the route falls within a high-risk area.

5.3.1.2 Societal Risk

Societal risk considers the overall likelihood of different events occurring. It can be much higher for transportation movements than is commonly found for fixed-facility operations, because of the multiplicative effect of route length and number of trips. In other cases, transportation risks can be much lower than facility risks if the route is remote from population, for example, some marine movements and pipelines. Figure 5.7 displays a typical F-N curve, which is the most common format for presenting societal risk.

Figure 5.7 Example F-N Curve

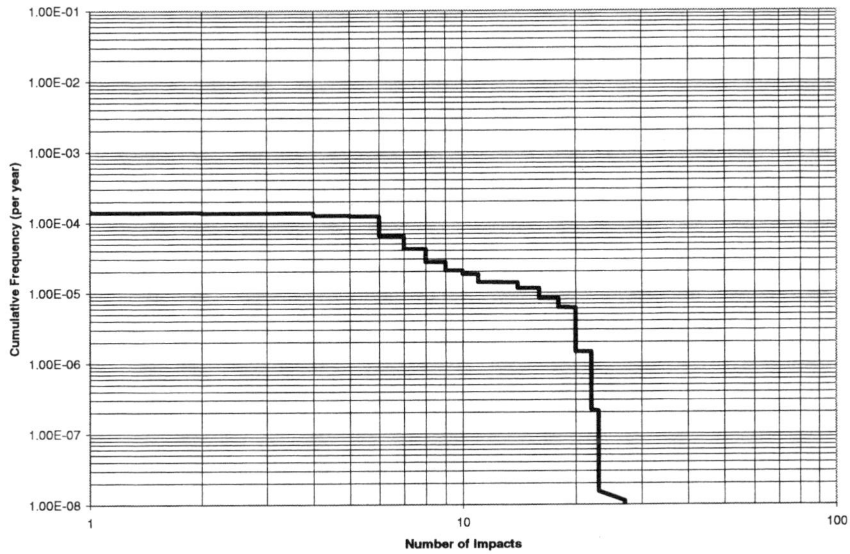

Societal risk is calculated to determine the likelihood of accidents involving multiple impacts. As shown in Figure 5.7, the F-N relationship is presented as a log-log plot with the x-axis representing the scale of the consequence in terms of N impacts and the y-axis representing F, the likelihood or expected frequency of N or more impacts. If available, an F-N curve can be plotted along with societal risk criteria for comparison, and to identify if risk reduction options need to be considered.

The F-N curve represents the cumulative frequency of all events exceeding a given outcome. To generate the F-N curve, the event probability and impact values (the F-N pairs for each scenario) from the risk analysis are sorted such that the largest impact is at the bottom of the table. For the largest impact, the event frequency value is the same as the cumulative frequency value. Next, the event frequency (associated with the next largest impact) is added to the event frequency (associated with the largest event) resulting in the cumulative frequency for this event. This process is repeated for all events to generate a table of cumulative frequency at various levels of impact, as illustrated in Table 5.2. The arrows indicate the sorting of the data to generate cumulative frequency and order of impacts. The resulting F-N curve using the data in Table 5.2 is presented in Figure 5.8.

Table 5.2 Example F-N Pair Data

Scenario ID	Frequency	Wind Direction	Wind Combination	Impacts	Cumulative Frequency
Node01_s_day	2.4E-02	S	D5	0	8.2E-02
Node01_s_day	3.1E-02	N	D5	1	5.8E-02
Node01_s_day	1.6E-02	E	D5	2	2.7E-02
Node01_s_day	7.8E-03	W	D5	3	1.1E-02
Node01_m_day	7.2E-04	S	D5	4	3.3E-03
Node01_m_day	2.4E-03	N	D5	7	2.6E-03
Node01_m_day	1.5E-05	W	D5	8	1.7E-04
Node01_m_day	1.5E-04	E	D5	10	1.6E-04
Node01_l_day	3.2E-06	S	D5	25	1.5E-05
Node01_l_day	6.5E-07	E	D5	35	1.2E-05
Node01_l_day	1.1E-05	N	D5	40	1.1E-05
Node01_l_day	6.5E-08	W	D5	50	6.5E-08

Figure 5.8 Example Resulting F-N Curve

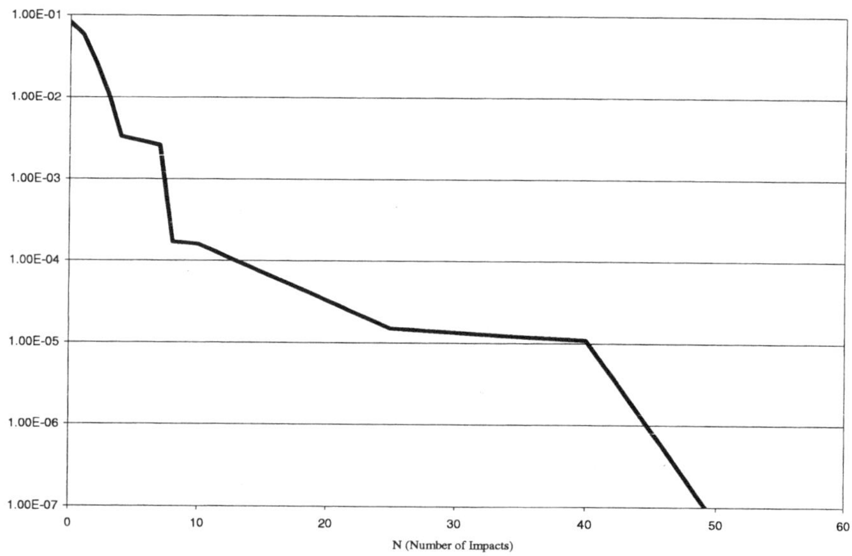

5.3.1.3 Risk Criteria

Risk criteria have been established by some regulatory agencies and companies for fixed-facility risks; however, there are very limited such target criteria for transportation risks, with a few companies beginning to use criteria for transportation risks. Where criteria are employed, initial risk estimates are compared against the criteria. If the risk target level is not met or is exceeded, the addition of risk reduction measures to reduce the risk estimate to (or below) the target risk level is evaluated.

There can be much greater disparity between individual and societal risk levels comparing fixed-facility to transportation. Therefore, internal calibrations of results based on past fixed-facility assessments should not be used.

5.3.1.4 Uncertainty

QRA results are determined using various likelihood databases, consequence models, onsite and offsite population data, and other assumptions. Each of these inputs has limited accuracy, therefore, there is uncertainty associated with risk assessment results. Various sources of uncertainty are found in risk calculations, and these sources can be classified according to the different levels of calculation (TNO, 1999):

- *Starting Points*—Before a risk analysis is started, choices must be made with regard to starting points. A conservative, worst-case approach requires a different type of model and different set of parameter values/assumptions than a best-estimate calculation.
- *Parameter Values*—As part of a risk assessment, parameters representing the inputs to the analysis are collected and developed. Sources of uncertainty may include, for example, impacts from exposure to a hazardous material (toxicity, thermal radiation, blast overpressure), physical data for the chemical, and accident data.
- *Models*—Uncertainties in models reflect the weaknesses, deficiencies, and inadequacies intrinsic to any model, and are a measure of the degree to which a model fails to represent reality.

Clearly, uncertainty can be an issue if the limitations of the inputs are not fully transparent or understood in the decision-making process. Therefore, as part of any risk assessment:

- All inputs and assumptions should be documented.
- Sensitivity analyses should be conducted to determine the influence of inputs.
- Risk estimates should not be treated as exact measurements, but as a best estimate of the risk level.

In the end, rather than solving problems, QRA results in and of themselves offer alternatives for management action and consideration. The greatest value is *not* in trying to establish an absolute measure of risk or in making decisions based on a single quantitative value. Rather, the greatest value is in providing a relative risk comparison (for the corporation, business, or operation) so that priorities for action can be set.

5.4 XYZ CHEMICAL EXAMPLE—QUANTITATIVE RISK ANALYSIS

As used in other chapters of this book, the XYZ Chemical example provides a simplified illustration of a QRA. An actual risk analysis may be more or less complex than what is discussed here. The details of the calculation are presented in Appendix E on the enclosed CD.

Through the use of qualitative and semi-quantitative risk analysis techniques, XYZ Chemical was able to prioritize, evaluate, and manage (or control) the majority of the risks across their operations worldwide. Following all of the individual facilitated team reviews (detailed in Chapter 4 for the Asian facility), a concern remains regarding tanker truck deliveries of pesticide that the company ships from a limited number of facilities worldwide. The concern arises because some of the current carrier routes pass through areas of significant population density and several of the carrier accident rates are high. If an accidental release occurred in one of these high-density locations, the following could result:

- Potential large scale evacuation (or shelter-in-place) of the residents in this area
- Potential for an unacceptable number of health impacts
- Potential damage to the company's reputation and image as a safe operator and responsible neighbor in the communities in which it operates

Due to the potential impacts and safety record of some of the carriers, corporate commissioned a QRA of all pesticide tank truck operations that were escalated as a higher-risk issue from the semi-quantitative risk analysis. The focus of this example is on a single origin-designation pair from the Asian facility.

5.4.1 Scope Definition

Following a corporate directive that a QRA be conducted for the pesticide transportation operations from the Asian facility, the transportation manager at the facility is contacted by the facility's health and safety manager since the facility ships tank trucks of pesticide, the current carrier uses a route that passes through an area of high population, and the accident rate for the carrier may potentially drive the risk of this operation to the higher level. The health and safety manager informs the transportation manager that the QRA will be led by corporate EHS in the United States and that a consultant has been brought in to conduct the QRAs for all the affected operations. This corporate-level ownership of the process will provide consistency in data, analysis techniques, and results comparison, and will also enable the development of a consistent set of risk mitigation recommendations. The transportation manager, as the person responsible for all materials in and out of the plant, including tanks trucks of pesticide, is asked to participate in the QRA activities.

During the QRA kick-off meeting with corporate EHS, local representatives (including the transportation manager), and the risk management consultant, the following steps of the QRA process are outlined:

- *Step 1*—selecting scenarios/issues for evaluation
- *Step 2*—collecting and evaluating data
- *Step 3*—selecting consequence measures
- *Step 4*—conducting the QRA
- *Step 5*—presenting results

Step 1 frames the project scope. Through discussions with the team, and to ensure consistency with the other QRAs being conducted simultaneously across the company, the following scope of the QRA is defined:

- Focus is on accidental releases of pesticide and the potential for impacts to people (could be defined as fatalities, injuries, evacuations, or a combination, but for the example all will be grouped as "potential impacts").
- Routes will be segmented to identify the corridors of various population densities.
- The consultant's analysis of representative population densities in the countries and specific areas of transit will be applied.
- The transportation manager will work with carriers and the consultant to collect carrier-specific accident data.
- Current route will be evaluated and compared to:
  - Option 1—new carrier and route (a greater total distance, but the new route travels through a relatively constant population density, described as suburban)
  - Option 2—current carrier with smaller tank trucks, but more frequent deliveries (current route)
- Results of the study will be in the form of individual and societal risk estimates.
- No formal risk criteria will be utilized at the facility level, and the assessment will illustrate only a comparison of the current operation to the alternatives.

Figure 5.9 illustrates the routes that will be analyzed as part of this QRA. As shown, the current route was segmented and passes through three different levels of population density:

- *Segment 3*—urban (high-population density)
- *Segments 1 and 4* —suburban (medium-population density)
- *Segments 2 and 5*—rural (low-population density)

Figure 5.9 Pesticide Tank Truck Routes

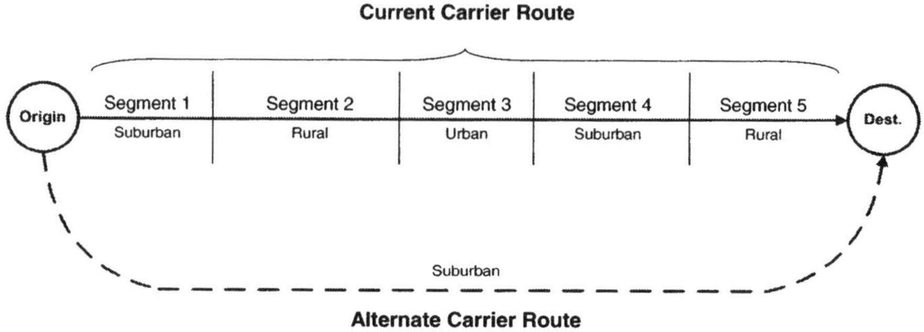

5.4.2 Data Collection

To complete the QRA, several sources of data must be collected. Table 5.3 details the data and responsibilities across the project team.

Table 5.3 QRA Data Needs—Pesticide Tank Truck

Data Category	Local Facility	Corporate EHS	Consultant
COMMODITY DATA:			
• Chemical Physical	X		
• Properties	X		
• Packaging Types	X		
• Annual Shipments			
ROUTE DATA:			
• Route Segments	X	X	X
• Meteorological Conditions			X
DEMOGRAPHIC DATA:			
• Population Densities		X	X
FREQUENCY DATA:			
• Carrier Accident Rates	X		
• Release Probabilities	X		X
• Spill Size Distribution			X
• Post-Incident Outcomes			X
CONSEQUENCE DATA:			
• Consequence Modeling		X	X
• Impact Criteria			X

As shown in Table 5.3, the local facility (and the transportation manager) is responsible for the following:

- All the commodity data, including the chemical properties of the pesticide, packaging types (including quantities shipped per transit), and annual shipment information including:

- Current Operation:
  - Shipment size: 10,000 gallons
  - Shipments: 200 shipments per year
  - Route length: 300 miles round trip (½ full, ½ empty)
- Option 1:
  - Shipment size: 10,000 gallons
  - Shipments: 200 shipments per year
  - Route length: 500 miles round trip (½ full, ½ empty)
- Option 2:
  - Shipment size: 5,000 gallons
  - Shipments: 400 shipments per year
  - Route length: 300 miles round trip (½ full, ½ empty)

- For the route data, the entire team contributed to the selection of route segments. The consultant is responsible for evaluating meteorological conditions (inputs to the consequence modeling) along the route. The route lengths and segments include:
  - Current Operation and Option 2:
    - 5 segments (total loaded distance = 150 miles):
      - Urban: 30 miles
      - Suburban: 60 miles
      - Rural: 60 miles
    - Option 1:
      - 1 segment (total loaded distance = 250 miles):
        - Suburban: 250 miles
- Demographic data defined by the consultant and approved by corporate EHS was used for the quantification of the operation's risks. For these routes the following are defined:
  - Urban: 3,000 people per square mile
  - Suburban: 1,000 people per square mile
  - Rural: 100 people per square mile
- Frequency data will be developed by the consultant, but the transportation manager will support this activity by working with the current and proposed carriers on their accident rate and release data. This activity is discussed below under frequency analysis.
- Consequence data will be the primary responsibility of the consultant, with corporate EHS defining the impact criteria for exposure to a pesticide release. Additionally, the impact criteria will be consistent across all QRAs conducted by XYZ Chemical.

5.4.3 Freqeuncy Analysis

As part of the support activities for the QRA, the transportation manager works with the current and proposed carriers reviewing accident statistics. Since the delivery of pesticides from the facility is a dedicated activity, it is important to develop representative accident rates, as opposed to using a general trucking accident rate. Both carriers keep excellent records for all their hazardous material transportation operations. The following is determined and provided as input to QRA after a review of these records:

- The current carrier had over 10 years of experience carrying pesticides in this area.
- The proposed carrier was new to this activity, but had been shipping other hazardous materials over a similar time frame to the existing carrier.
- The consultant determined that both carriers use the same tank truck specification so the conditional release probabilities and spill size distributions would be constant for both options. Additionally, Option 2 (use of a smaller tank truck) would be of a similar design to the larger tank trucks, so the conditional probabilities are equal for the current operation and the two options.
- The carrier datasets were not detailed enough to generate segment-specific accident rates, but overall carrier specific accident rates were developed:
  - Current carrier accident rate: 1.5 accidents per million miles
  - Proposed carrier accident rate: 2.5 accidents per million miles

As calculated, the proposed carrier's accident rate is 1.7 times higher than the current carrier. Based on this information, Table 5.4 shows the intermediate statistics:

5.4.4 Consequence Analysis

The consequence analysis portion of the QRA is completed by the consultant and the following data are provided for review:

- Three release sizes were defined (small, medium, and large). These release sizes dictated the release size and duration in the consequence analysis and the probability of the size of release in the frequency analysis.
- A single set of meteorological conditions was applied since a relatively small distance is traveled and all within the same area.
- Two wind speed/stability class combinations were used for modeling:
  - F Stability, 1.5 meters/second wind speed
  - D Stability, 5 meters/second wind speed

From these assumptions, Table 5.5 details the consequence modeling results provided by the consultant.

Table 5.4 Intermediate Accident Statistics (only for example purposes)

Data/Statistic	Current Operation	Option 1	Option 2
Accident Rate (accidents/mile)	1.5E-06	2.5E-06	1.5E-06
Annual Loaded Miles (shipments x loaded miles)	30,000	50,000	60,000
Estimated Accidents (accident rate x loaded miles)	0.045 per year	0.125 per year	0.09 per year
Accident Recurrence Period (1/estimated accidents)	22 years	8 years	11 years

Table 5.5 Example Hazard Zones—Pesticide Release

Option	Scenario (Release Size)	Stability / Wind Speed	Toxic Dispersion	
			Distance (meters)	Area of Impact (m^2)
Current (10,000 gallons)	Small Medium Large	D / 5 m/s F / 1.5 m/s D / 5 m/s F / 1.5 m/s D / 5 m/s F / 1.5 m/s	50 100 150 300 250 500	300 1,000 3,000 10,000 8,000 30,000
Option 1 (10,000 gallons)	Small Medium Large	D / 5 m/s F / 1.5 m/s D / 5m/s F / 1.5 m/s D / 5 m/s F / 1.5 m/s	50 100 150 300 250 500	300 1,000 3,000 10,000 8,000 30,000
Option 2 (5,000 gallons)	Small Medium Large	D / 5 m/s F / 1.5 m/s D / 5 m/s F / 1.5 m/s D / 5 m/s F / 1.5 m/s	25 50 75 150 100 175	100 300 600 3,000 1,000 4,000

Since the only potential outcome of concern for the QRA is a toxic release and impacts to people surrounding the route, an event tree is not required for this analysis.

5.4.5 Risk Estimation and Discussion of Results

The results of the QRA are presented for the current pesticide transportation operation and the two options. To review, the options being considered include:

- *Option 1*—new carrier and route
- *Option 2*—current carrier with smaller tank trucks, but more frequent deliveries (current route)

The results that have been calculated for this operation include:

- Average Individual Risk
  - Current operation: 1.9E-06 per year
  - Option 1: 6.1E-07 per year
  - Option 2: 1.0E-06 per year
  - Option1 has the lowest individual risk, which is driven by the fact that the entire route runs through a suburban population density as compared to the full range of population densities for the current operation and Option 2. Additionally, the individual risk for Option 2 is lower than the current option due to the shorter dispersion distances, which is attributable to the smaller package design (as detailed in Table 5.5).
- Societal Risk
  - Societal risk is presented in the form of an F-N curve. The F-N curve comparing the pesticide transportation options is displayed in Figure 5.10, which illustrates the following:
    - The current transportation route poses the highest risk, in terms of both maximum estimated number of impacts and the frequency of occurrences.
    - Option 1 has a maximum number of impacts 67% lower than the current operation. Additionally there is a slight reduction in the likelihood at the lower-impact levels.
    - Option 2 has a maximum number of impacts 87% lower than the current operation and 60% lower than Option 1. Option 2 has the largest reduction in likelihood at the lower-impact levels.
- Average Rate of Impact
  - The average rate of impact is the sum of the F-N pairs, which was calculated for each option:
    - Current operation: 1.8E-01 impacts/year (1 impact every 6 years)
    - Option 1: 9.4E-02 impacts/year (1 impact every 11 years)
    - Option 2: 3.2E-02 impacts/year (1 impact every 30 years)

From these results, Option 1 (New Carrier) has the lowest individual risk, but Option 2 (Current Carrier—Smaller Shipment Size) has the lowest societal risk and average rate of impact. Option 2 illustrates that the risk reduction is driven by the lower potential consequences. This reduction is significant since this level of risk reduction was achieved even with doubling the annual mileage.

Since risk criteria were not defined at the operations level in this example, the results are sent back to management for comparison to the other facilities transporting pesticide by tank truck. One of the valuable insights from this example is that rerouting through a less populated area reduced the risk slightly, but still may not meet the risk reduction desired by management. Simply working with the carrier to manage the quantity in each shipment showed the greatest risk reduction benefits in this example, and these benefits will be considered further in Chapter 7 (including the cost-benefit of these options).

Figure 5.10 F-N Curve Pesticide Route Option Comparison

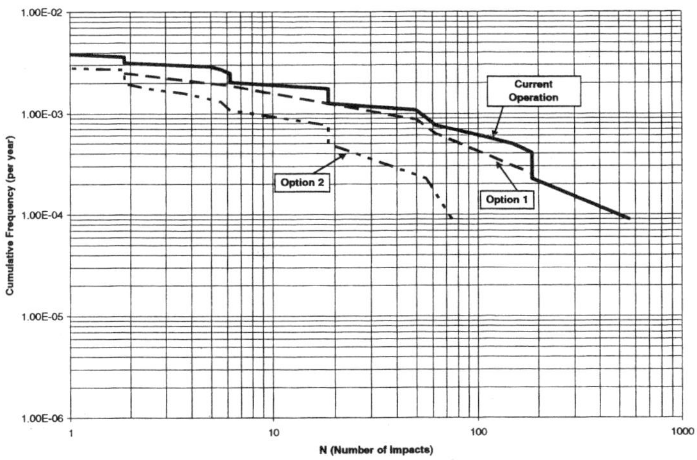

REFERENCES

Center for Chemical Process Safety, *Guidelines for Chemical Process Quantitative Risk Analysis*, American Institute of Chemical Engineers, New York, 1989

Center for Chemical Process Safety, *Guidelines for Chemical Process Quantitative Risk Analysis, Second Edition*, American Institute of Chemical Engineers, New York, 2000

Center for Chemical Process Safety, *Guidelines for Chemical Transportation Risk Analysis,* American Institute of Chemical Engineers, New York, 1995

Center for Chemical Process Safety, *Guidelines for Consequence Analysis of Chemical Releases*, New York, 1999

TNO, *Guidelines for Quantitative Risk Assessment, First Edition,* CPR 18E, The Netherlands Organization of Applied Scientific Research, The Hague, 1999

TNO, *Methods for the Determination of Possible Damage, First Edition*, CPR 16E, The Netherlands Organization of Applied Scientific Research, The Hague, 1992

6
TRANSPORTATION SECURITY CONSIDERATIONS

6.1 OVERVIEW OF TRANSPORTATION SECURITY

Worldwide, there is an increased level of security awareness in all aspects of our daily activities. Specific to transportation security, we are faced with many challenges as a result of a global economy and supply chain. These challenges include not only terrorism, but also piracy, organized crime, illegal drug manufacturing and trafficking, and smuggling. It is important to consider all potential threats when evaluating chemical transportation security, since a chemical shipment can fall victim to a number of different types of attacks, not just terrorism. The good news is that many of the same security countermeasures can be effective in addressing these widely divergent security threats.

A number of terrorist-related attacks have highlighted the vulnerabilities of the transportation infrastructure. Examples of some significant attacks against, or through the use of, transportation assets are listed in Table 6.1. These types of attacks, combined with recent accidents, are changing the perception, tolerance, and means of managing the risk of hazardous material transportation operations. It is increasingly common that strong public reaction following a hazardous material transportation accident is not limited to safety, but also focuses on transportation security. Specific accidents in the United States include the following

- *Graniteville, South Carolina (January 6. 2005)*—train derailment and chlorine release
- *Minot, North Dakota (January 18, 2002)*—train derailment and anhydrous ammonia release
- *San Antonio, Texas (June 28, 2004)*—train derailment and chlorine release

These incidents have highlighted the vulnerabilities of the transportation infrastructure and the potential consequences if a shipment was compromised en route. This concern has led to worldwide calls for legislation directed at increased security, rerouting of hazardous material shipments, as well as a push for the application of inherently safer technologies (commonly referred to as IST).

Table 6.1 Examples of Attacks Against Transportation Infrastructure

Event	Infra-structure	Description
London Bombings (7 July 2005)	Public Transit (Road and Rail)	The 7 July 2005 London bombings were a series of coordinated terrorist bomb blasts that hit London's public transport system during the morning rush hour. Three bombs exploded on three London underground trains while a fourth bomb exploded on a bus. The bombings killed 52 commuters and the four suicide bombers, injured 700, and caused a severe daylong disruption of the city's transport and mobile telecommunications infrastructure.
Madrid Spain Bombings (11 March 2004)	Public Transit (Rail)	The 2004 Madrid train incident consisted of a series of coordinated bombings against the Cercanías (commuter train) on the morning of 11 March 2004, killing 191 people and wounding approximately 2,000.
9/11 (11 September 2001)	Commercial Airlines (Air)	The attacks of 9/11 in the United States consisted of the hijacking of four commercial airliners resulting in the destruction of the two World Trade Center towers, damage to the Pentagon, and responsible for nearly 3,000 fatalities.
USS Cole (12 October 2000)	Military (Marine)	This attack was a suicide bombing attack against the U.S. Navy destroyer USS Cole on 12 October 2000, while it was anchored in the Yemeni port of Aden. Seventeen sailors were killed in the attack, which is often seen as a precursor to the September 11 terrorist attacks on U.S. soil.
Sarin Gas Attack on the Tokyo Subway (20 March 1995)	Public Transit (Rail)	The 20 March 1995 attack was an act of domestic terrorism in Japan perpetrated by members of Aum Shinrikyo (a Japanese religious cult). In five coordinated attacks, the conspirators released sarin gas on several lines of the Tokyo Metro, killing 12 people, severely injuring 50, and causing temporary vision problems for nearly a thousand others.
Iraq War	Theft (Road)	Chlorine reportedly has been used recently in bombings in Iraq with the intent of inciting additional fear and chaos from a secondary chlorine exposure.
General Attacks	(Pipeline)	Worldwide, there have been numerous reported attacks on oil and gas pipelines with the intent to create economic disruptions.

The reason for this increased focus on transportation security is the fear that hazardous materials could be attractive terrorist targets due to the potential consequences resulting from an intentional release, contamination of the material in transit, or theft of the material for use as a weapon against another target. Since the chemical industry is critical to the global economy, any successful attack, damage, or disruption could possibly cause negative local/regional/national impacts or rippling economic disruptions across numerous industry sectors. Therefore, the chemical industry and its value chain have been identified as a critical infrastructure requiring security measures in most countries to address today's threat environment.

In addition to these highly publicized attacks against the transportation infrastructure, the geographic diversity of the threat, the potential to sabotage of transportation infrastructure, and theft are also concerns. Besides the Madrid and London bombing listed in Table 6.1, the rail network has been attacked worldwide, including several train bombings in India (e.g., Mumbai train bombings on 11 July 2006) and Russia (Moscow metro bombing on 6 February 2004). Sabotage attacks to bridges and rail lines include the Moscow passenger train bombing (14 August 2007), Rafiganj India train derailment (10 September 2002), Jaunpur India train crash (13 May 2002), Kurnool India train crash (21 December 2002), and the Palo Verde, Arizona, train derailment (9 October 1995). Attacks on commercial freight shipping have also included the direct attack and bombing of the French oil tanker Limburg off the coast of Yemen (6 October 2002), the piracy of the Danica White off the coast of Somalia (1 June 2007), and the piracy of the supertanker Dewi Madrim in the Straits of Malacca (26 March 2003). Beyond direct attacks, there have been reported cases of smuggling of suicide bombers into Ashdod Port, Israel (14 March 2004), as well as the theft of hazardous materials from road shipments. Although presently theft is primarily the result of organized crime, as opposed to terrorism, there are established links between the money generated by certain organized crime elements and the funding of terrorist activities.

This chapter provides a discussion of current security guidelines and regulations (Section 6.2) and a methodology for adapting and applying security vulnerability assessment (SVA) techniques designed for fixed chemical facilities to account for the differences specific to transportation (Sections 6.3 and 6.4). Practical transportation security elements are provided in Section 6.5. Finally, at the end of the chapter, the XYZ Chemical example highlights the unique issues of a security vulnerability assessment as compared to a safety risk analysis.

6.2 TRANSPORTATION SECURITY CONCEPTS

This section outlines the basic information needed for conducting or participating in a transportation security vulnerability assessment (TSVA). A protocol for prioritizing security issues and a methodology for conducting a more detailed TSVA of escalated security issues are described in Sections 6.3 and 6.4, respectively. The purpose of this section is to provide a background and resources for:

- Security risk definitions
- Current security regulations
- Safety and security synergies and tradeoffs

6.2.1 Security Risk Definitions

The definitions presented here are based on the concepts detailed in the *Guidelines for Analyzing and Managing the Security Vulnerabilities of Fixed Chemical Sites* (CCPS,

2003). Individuals participating in a general security review, security prioritization activity, or TSVA should review and be familiar with the following definitions:

- *Adversary*—any individual, group, organization, or government that conducts activities, or has the intention and capability to conduct activities detrimental to critical assets. An adversary can include intelligence services of host nations, or third-party nations, political and terrorist groups, criminals, disgruntled employees or contractors, and private interests. Adversaries also can include insiders, outsiders, or the two acting in collusion.
- *Alert Levels*—a progressive, qualitative measure of the likelihood of terrorist actions, from negligible to imminent, based on government or company intelligence information. Different security measures may be implemented at each alert level based on the level of threat.
- *Asset*—any person, environment, facility, material, information, business reputation, or activity that has a positive value to an owner. The asset may have value to an adversary, as well as an owner, although the nature and magnitude of those values may differ.
- *Asset Category*—among these are:
  - People
  - Chemicals
  - Information
  - Environment
  - Equipment
  - Facilities
  - Activities/operations
- *Capability*—when assessing the capability of an adversary, two distinct capabilities need to be considered. The first is the capability to obtain, damage, or destroy the asset. The second is the adversary's capability to use the asset to achieve its objectives once the asset is obtained, damaged, or destroyed.
- *Countermeasures*—actions taken or a physical capability provided whose principal purposes are to reduce or eliminate one or more vulnerabilities, for example, an effective emergency response, or the intent or capability of the adversary
- *Critical Infrastructure*—systems and assets, whether physical or virtual, so vital that the loss, interruption, incapacity, or destruction of which (1) would have a negative or debilitating effect on the security, economic security, public health, or safety of a nation, region, or any local government or (2) cause national or regional catastrophic effects.
- *Delay*—a countermeasure strategy that provides various barriers to slow the progress of an adversary to prevent an attack or theft, to increase the time necessary to cause an event, or to assist in apprehension and prevention of theft

- *Detection*—a countermeasure strategy intended to identify an adversary attempting to plan or commit a chemical security event or other criminal activity via either real-time observations or pre-incident analysis of the activities/intelligence
- *Deterrence*—a countermeasure strategy intended to prevent or discourage the occurrence of a breach of security by means of fear or doubt
- *Intent*—a course of action an adversary intends to follow. When assessing threats, security professionals need to evaluate intent as well as capabilities. To determine the intent and motivation of an adversary, an adversary's goals and objectives, as well as specific events that might trigger an adversary's act, must be closely examined. The questions that should be asked about intent are: "Does the adversary have a current or projected need for this asset? Does the adversary seek to deny or destroy the use of this asset?"
- *Layers of Protection*—a concept whereby several independent devices, systems, or actions are provided to reduce the likelihood and severity of an undesired event
- *Likelihood of Adversary Success (L_{AS})*—the potential for causing a catastrophic event by defeating the countermeasures. L_{AS} is an estimate whether the security countermeasures will thwart or withstand the attempted attack or whether the attack will circumvent or exceed the existing security measures. This measure represents a surrogate for the conditional probability of success of the event.
- *Response*—a security strategy to neutralize the adversary or to evacuate, shelter in place, call local authorities, control a release, or take other mitigation actions
- *Risk*—the potential for damage to, or loss of, an asset. Risk, in the context of chemical process security, is the potential for the intentional event outcome to be realized. Examples typically of interest include an intentional release of hazardous materials from containment, the theft of chemicals that could later be used as weapons, the contamination of chemicals that may later harm the public, the economic costs of the damage, or disruption of the chemical process or other nearby critical infrastructure. Therefore, risk is an expression of the likelihood (L_{AS}) that a specific vulnerability (V) of a particular attractive target (A_T) will be exploited by a defined threat (T) to cause a given consequence (C).
- *Security Plan*—a document that describes a plan to address security issues and related events including security assessment and mitigation options. This includes security alert levels and response measures to security threats.
- *Security Vulnerability Analysis (SVA)*—the process of determining the likelihood of an adversary successfully exploiting a vulnerability and the resulting degree of damage or impact. An SVA is not a quantitative risk analysis, but is performed qualitatively using a structured and repeatable method and the best judgment of safety, security, and transportation personnel. A qualitative determination of risk is the desired outcome of an

SVA in that it provides the basis for ranking the security-related risks and thus establishing priorities for the application of countermeasures.

- *Transportation Security Vulnerability Analysis (TSVA)*—modified SVA process specifically designed to account for specific threats, consequences, vulnerabilities, risks, and security countermeasures of transportation activities, including hazardous materials in transit
- *Target Attractiveness*—an estimate of the value of a target to an adversary based on the factors shown below. Experience has shown that, particularly for terrorist attacks, certain targets better accomplish the objectives of the adversaries than others. Consideration should be given to these factors in defining the threat and in determining the need for any enhanced countermeasures:
  - Potential for mass casualties/fatalities
  - Extensive property damage
  - Proximity to national assets or landmarks
  - Possible disruption or damage to critical infrastructure
  - Disruption of the national, regional, or local economy
  - Ease of access to target
  - Media attention or possible interest of the media
  - Company reputation and brand exposure
- *Terrorism*—defined in the Homeland Security Act of 2002 as any activity that (A) involves an act that (i) is dangerous to human life or potentially destructive of critical infrastructure or key resources; and (ii) is a violation of the criminal laws of the United States or of any State or other subdivision of the United States; and (B) appears to be intended (i) to intimidate or coerce a civilian population; (ii) to influence the policy of a government by intimidation or coercion; or (iii) to affect the conduct of a government by mass destruction, assassination, or kidnapping.
- *Threat*—any indication, circumstance, or event with the potential to cause the loss of, or damage to, an asset. A threat can also be defined as the intention and capability of an adversary to undertake actions that would be detrimental to critical assets.
- *Vulnerability*—any weakness that can be exploited by an adversary to gain access to an asset

6.2.2 Security Regulations and Industry Practices

In response to the potential intentional attacks targeting transportation infrastructure, regulations and industry guidance have been developed and implemented to enhance the security while transporting hazardous materials. While a detailed review of transportation regulations was presented in Chapter 2, this section provides a resource and summary to the different regulatory sources, international applicability, cross-border security initiatives, and a summary of current requirements. Appendix C, on the enclosed CD, can be used as a resource for specific regulations in other countries and emerging markets. The use of these links and other sources will help ensure that the most current information on transportation security can be easily referenced.

6.2.2.1 International Security Regulations

The UN Model Regulations are the primary source for international regulations pertaining to the transport of hazardous materials (or dangerous goods). As a good starting point for determining specific security requirements by country and mode of transit, Table 6.2 presents a list of resources for international transportation regulations covering the following:

- General security provisions
- Security training
- Additional requirements for high-consequence dangerous goods
- Security plans

Table 6.2 Sources of International Security Regulations

Applicability	Reference
General Source for International Regulations	UNEC United Nations Economic Commission for Europe: (http://www.unece.org/Welcome.html)
Road	European Agreement concerning the International Agreement of Dangerous Goods by Road (ADR): (http://www.unece.org/trans/danger/publi/adr/adr_e.html)
Rail	Intergovernmental Organization for International Carriage by Rail (OTIF): (http://www.otif.org)
Marine	International Maritime Organization: (http://www.imo.org)
Air	International Civil Aviation Organization (ICAO): (http://www.icao.int)

6.2.2.2 U.S. Security Regulations

The Pipeline and Hazardous Materials Safety Administration (PHMSA), a part of the U.S. Department of Transportation (U.S. DOT), is responsible for the safe and secure movement of hazardous materials by all transportation modes. The primary office within PHMSA responsible for transportation safety and security is the Office of Hazardous Materials Safety. In addition to PHMSA, the Federal Railroad Administration and the Federal Motor Carrier Safety Administration within U.S. DOT are involved in transportation safety and security.

A number of regulations implemented by various agencies cover many aspects of transportation security. The regulation that affects the majority of regulated sites is PHMSA's *Hazardous Materials: Security Requirements for Offerors and Transporters of Hazardous Materials* (49 CFR Part 172, Subpart I). This regulation, which became effective in 2003, requires any offeror or transporter of certain hazardous materials to develop a security plan (including sections on personnel security, unauthorized access, and en-route security) and to conduct security training.

The Maritime Transportation Security Act of 2002 (MTSA) required the establishment of security regulations to increase the security of the nation's ports. These regulations are published under 33 CFR 101 (Maritime Security: General), 33 CFR 103 (Maritime, Security: Area Maritime Security), 33 CFR 104 (Maritime Security: Vessels), 33 CFR 105 (Maritime Security: Facilities), and 33 CFR 106 (Maritime Security: Outer Continental Shelf Facilities), and are enforced by the U.S. Coast Guard. The purpose of these regulations is to:

- Align requirements of domestic maritime security regulations with the international maritime security standards in the International Convention for Safety of Life at Sea, 1974 (SOLAS Chapter XI-2) and the International Code for the Security of Ships and of Port Facilities, parts A and B, adopted on 12 December 2002;
- Ensure that security arrangements are as compatible as possible for vessels trading internationally;
- Emphasize cooperation and coordination with local port community stakeholders, based on existing domestic standards as well as established industry security practices; and
- Require assessments and plans intended for use in implementing security measures at various maritime security (MARSEC) levels.

The Department of Homeland Security (DHS) published the interim final rule of the Chemical Facility Anti-Terrorism Standards (CFATS) (6 CFR. 27, April 9, 2007), giving DHS the authority to regulate "high-risk" chemical facilities. While this regulation does not currently apply to hazardous materials in transit, or facilities outside of the United States, it signifies a heightened priority for enhancing security across the chemical industry. In addition to DHS, other agencies, including the Transportation Security Administration (TSA), the U.S. Coast Guard (USCG), and Customs and Border Protection (CBP) have been given new security directives, and have proposed their own security regulations for hazardous material shipments. These agencies should be monitored for new security requirements in the future, with Table 6.3 providing a list of resources for United States government agencies with security responsibilities.

Table 6.3 Federal Government Resources

Reference	Link
U.S. Department of Transportation (DOT) Pipeline and Hazardous Materials Administration Office of Hazardous Materials Safety	http://www.dot.gov/ http://www.phmsa.dot.gov/ http://hazmat.dot.gov/
U.S. Department of Homeland Security	http://www.dhs.gov/
U.S. Coast Guard	http://www.uscg.mil/
U.S. Customs and Border Protection	http://www.cbp.gov/
Transportation Security Administration	http://www.tsa.gov/
Federal Railroad Administration	http://www.fra.dot.gov/
Federal Motor Carrier Safety Administration	http://www.fmcsa.dot.gov/

6.2.2.3 Cross-Border Initiatives

In November 2001, CBP established the Customs-Trade Partnership Against Terrorism (C-TPAT), a voluntary means for driving improvements in the security of the importation process. (The Safe Port Act of 2006 transformed the elements of this voluntary program into law). The intent of the program is to ensure protection of the supply chain from shipment origin outside of the U.S. borders to the final destination in the United States. By complying with the required security procedures and practices, participating U.S. companies receive benefits during the importation process. Chief among these benefits is a streamlined Customs process at the ports and border crossings that can significantly shorten shipment transit (hold) times and thus provide "C-TPAT approved" entities a competitive or cost advantage. Procedures include the establishment of:

- Procedural security controls
- Physical security controls
- Access controls
- Personnel security
- Conveyance security

Where an importer does not exercise physical control over a facility that supplies the product, or the means by which the product is transported into the United States, the importer is expected to ensure that its supply chain partners agree to comply with the required security components of the C-TPAT program. The C-TPAT program is the most internationally far-reaching transportation security initiative as it extends beyond the U.S. borders and directly involves foreign supply chain partners. As this program continues to evolve, companies should be aware of similar international initiatives, which could further increase requirements and specific security requirements.

The World Customs Organization (WCO) is extending the C-TPAT initiative worldwide. In June 2005, the WCO adopted the Framework of Standards to Secure and Facilitate Global Trade (SAFE), an international instrument containing 17 standards that promote security, trade facilitation, anticorruption, and revenue collection. The SAFE Framework aims to:

- Establish standards that provide supply chain security and facilitation at a global level to promote certainty and predictability
- Enable integrated supply chain management for all modes of transport
- Enhance the role, functions, and capabilities of Customs to meet the challenges and opportunities of the twenty-first century
- Strengthen cooperation between international customs administrations to improve their capability to detect high-risk consignments
- Strengthen Customs/business cooperation
- Promote the seamless movement of goods through secure international trade supply chains

Additionally, European Union (EU) directive EUD 648/2005 went into effect on 1 January 2008 thus permitting Authorized Economic Operators (AEO) to receive a certificate for safety and security. Certified companies will have a reduction on cross-border checks, and non-certified companies may see an increase in number and depth of checks.

6.2.2.4 Industry Guidance

Industry guidance assists in meeting regulatory requirements and increasing the security of hazardous materials in transit. Specific to transportation security, the following guidance documents are good resources on transportation security, security management systems, and vulnerability assessment techniques:

- American Chemistry Council Responsible® Care Security Code of Management Practices
- *Implementation Resource Guide for Responsible Care® Security Code of Management Practices: Value Chain Activities*, American Chemistry Council, 2002
- *Hazardous Materials Transportation Enhanced Security Requirements*, U.S. Department of Transportation Research and Special Programs Administration

Some industry, chemical, and transportation association resources and links are listed in Table 6.4.

Table 6.4 U.S. Industry Security Resources

Reference	Link
American Chemistry Council (ACC)	www.americanchemistry.com
American Petroleum Institute (API)	www.api.org
American Society for Industrial Security	www.asisonline.org
American Trucking Associations	www.truckonline.com
Association of American Railroads (AAR)	www.aar.org
Center for Chemical Process Safety, American Institute of Chemical Engineers	www.aiche.org/ccps
Chlorine Institute	www.cl2.com
Compressed Gas Association	www.cganet.com
The Fertilizer Institute	www.tfi.org
Institute of Makers of Explosives	www.ime.org
National Association of Chemical Distributors	www.nacd.com
National Propane Gas Association	www.npga.org
National Tank Truck Carriers	www.tanktransport.com
Security Industry Association	www.siaonline.org
Synthetic Organic Chemical Manufacturers Association (SOCMA)	www.socma.com

6.2.3 Safety and Security Synergies and Tradeoffs

When developing or applying baseline security elements and proposing recommended countermeasures as part of a security review, one must understand compromises between safety, security, and convenience. The key difference between security and safety is that potential security incidents are the result of intentional acts. This is the basis for understanding the hazards, potential consequences, and the likelihood of an intentional malevolent event. While an increase in security may result in changes and inconveniences that everyone in the supply chain will need to understand, accept, and make appropriate adjustments, security decisions should not be blindly made without considering transportation safety implications. As such, it is recommended that security be integrated into the overall transportation risk management process. Examples of safety and security synergies and tradeoffs are presented in Table 6.5.

Table 6.5 Examples of Safety and Security Synergies and Tradeoffs

Issue	Safety	Security
Placards	Commodity information needed by emergency responders to react appropriately to an accident and minimize any impact.	Commodity information could be used by terrorists to target specific chemicals.
Rerouting	May result in more accidents if there are longer transits or the infrastructure along an alternate route may be less well maintained or contain undesirable features (uncontrolled intersections, no shoulders, etc.).	Eliminating a shipment near a specific location (most likely a highly populated or critical area) may inadvertently transfer the risk from one community to another.
Working with supply chain partners (implementing security countermeasures)	Technology can be used for both safety and security (e.g., GPS to indicate location en route, emergency response to accident, and monitoring time-sensitive chemicals/materials).	Technologies focused on security should not distract the main function of the carriers (e.g., the safe transport of chemicals from point A to B).
Risk Analysis Methods	• Rational and structured results lead to recommendations • Participation and engagement by individuals with different perspectives, roles, and backgrounds/skill sets for safety, security, and transportation • Similar methodology • Same decision metrics (guidelines)	

6.3 SECURITY PRIORITIZATION PROCESS

Similar to the safety risk analyses, a security screening process should be utilized to identify specific issues that require a full TSVA, screening out those scenarios that justify only a general security review. As compared to the prioritization process focused on safety, security screening brings about new concerns; therefore, the chemicals and movements escalated through each distinct activity may be different. Some chemicals and movements that may be ranked as a lower priority, or even screened out for further detailed analysis based on safety issues, may rank *higher* in light of the security concerns. For example:

- A material transported in a specially designed container to minimize the likelihood of a release based on safety concerns (accidental release) may have ranked low as having a potential for a large release. However, since the goal of an intentional act is to create a maximum release, a larger security event may be possible than that anticipated from an accidental scenario.
- The transportation of a chemical in a limited number of small cylinders may have ranked low from a safety concern. However, this activity might be escalated for security reasons based on the potential for theft and the subsequent use of this material to produce a chemical weapon that could be mobilized and used against any number of potential targets.
- A food additive may have screened out of a safety analysis since there are no hazards associated with the product, but the potential for tampering with a loaded tank truck or rail car may raise the concern of potential contamination of the product. Therefore, a review of the measures in place to prevent and detect tampering at the origin, along the route, and at the destination may be warranted.
- A tank truck of flammables may have ranked low in a safety analysis as the consequence footprint of such materials can be much smaller than that of toxics. But, if an attack along this route has the ability to result in a consequence that could significantly damage a critical transportation route and result in an extended shutdown for cleanup and repair, then this segment of the route may warrant escalation in the security review. Also, the shipment could be hijacked, enabling the flammables to be diverted from the planned route.

Specifically for security, the following factors may be helpful in developing a company-specific security prioritization process:

- Chemical hazards
- Quantity transported per container
- Number of shipments or campaigns
- Mode of transit
- Proximity to people
- Proximity to sensitive environmental areas
- Interim (unsecured) storage
- Proximity to other critical assets or infrastructure
- Specific threat information

No one prioritization factor is intended to escalate an issue to a full TSVA; however, it is recommended that a multistep process be developed to identify the specific chemicals, movements, and segments that may require further evaluation. An example prioritization process is illustrated in Figure 6.1, consisting of the following steps:

- *Step 1—List Chemicals and Hazards*: The inherent physical properties (e.g., toxicity, flammability, reactivity) of a hazardous material in transit should be used to develop a list of chemicals that will be carried forward in the prioritization process. This list of chemicals may be different than that developed for accidental releases and may need to consider the following security issues:
  - Loss of significant containment of a hazardous material in transit from intentional damage of equipment or malicious release with the potential to impact people, property, or the environment
  - Theft, diversion, and misuse with the intent to cause harm along the route or an alternate location
  - Contamination with the intent to cause a reactive event while in transit, an event at the final destination, or other offsite impacts that could result from tampering with the material
- *Step 2—Review Mode(s) and Quantities Shipped*: For all chemicals carried forward from Step 1, the mode and quantities should be reviewed to further refine a list of chemicals and modes of transportation and may be based on:
  - Minimum quantity or a threshold for toxics, flammables, and explosives that, if released in transit, could pose an impact to people, property, or the environment
  - Dedicated or contract carrier versus common carrier
  - Features of the mode related to the difficulty of attack or vulnerability during various stages of the operation or at specific segments along the route
- *Step 3—Select Route or Movement for Evaluation*: Depending upon the complexity of the supply chain network, it may first be desirable to divide the transportation activities into more manageable groups. These may include specific chemicals or chemical classes, modes of transportation, movements, regions, in-transit storage (yard or warehouse), etc.
- *Step 4—Identify Sensitive Areas Along the Route*: Except for chemicals that remain on the list because of potential theft and contamination issues, the chemicals and modes carried forward from Step 3 should be evaluated to focus on specific sensitive segments along the route, which may include:
  - Proximity to people
  - Proximity to sensitive environmental areas
  - Proximity to other critical assets or infrastructure
  - Attack along the route that could result in the degradation of critical infrastructure or other critical assets

- *Step 5—Security Evaluation*: Each of the chemical movements screened out through the prioritization process should be analyzed by a general security review before closing these items. Section 6.5 discusses baseline security elements. All issues that make it through the prioritization process should be reviewed using a TSVA methodology (Section 6.4).
- *Step 6—Periodic Review*: It is important to periodically review the security of the transportation network to ensure that the baseline program elements are still in place, recommended countermeasures have been implemented, and vulnerabilities reassessed. Additionally, this process should be reinitiated if there is specific intelligence or threat information pertaining to your supply chain, mode of transport, or specific chemical movement.

Figure 6.1 Security Prioritization Process

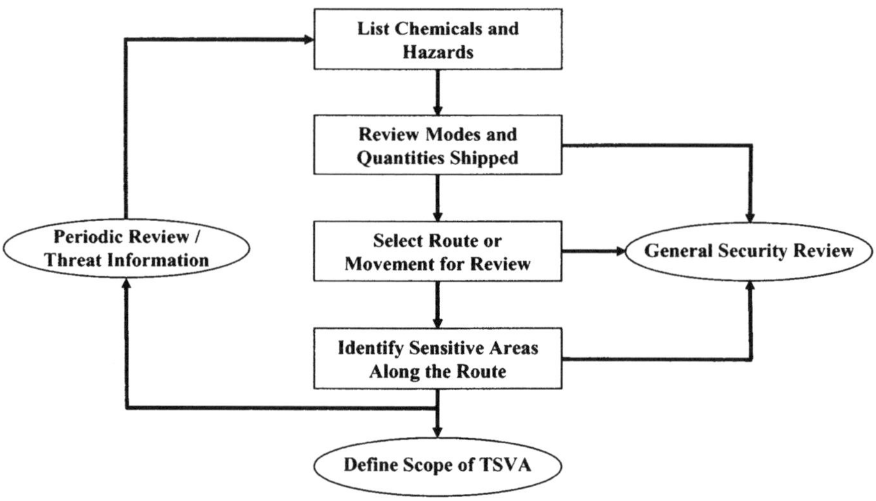

Following this prioritization process or a similar one, a transportation system can be evaluated to identify those issues that need to be escalated to a more detailed TSVA to better understand the potential consequences, vulnerabilities, and level of risk, or to compare potential security countermeasures.

6.4 TRANSPORTATION SECURITY VULNERABILITY ASSESSMENT

A TSVA can be designed to analyze security risk from a wide range of threats, from vandalism to terrorism. In theory, risk is a function of the consequence and likelihood of the range of credible transportation threat scenarios. For accidental risk analyses, consequence and likelihood are evaluated directly, but in security analyses, likelihood is evaluated as a function of vulnerability and threat. Therefore, security risk is defined as a measure of the consequences, vulnerabilities, and threats:

$$\text{Risk} = f(\text{C, V, T})$$

6.4.1 Consequence

The consequences of a security event at a chemical facility are generally expressed in terms of impacts to people (e.g., fatality, injury), property damage, and environmental damage. This definition of consequences is the same as that used for accidental releases and is appropriate for security-related transportation events. The key difference is that security-related events may involve more severe effects than those expected with accidental risk. Some relevant consequences in TSVAs include:

- Public fatalities and injuries
- Transportation personnel fatalities and injuries
- On-site fatalities and injuries at product origin and destination
- Emergency responder fatalities and injuries (e.g., provoking an initial incident to trigger a response, then a second strike aimed at the response)
- Theft of hazardous materials
- Disruption of the local, regional, or national economy
- Disruption of company operations
- Environmental damage
- Financial loss
- Secondary damage to critical infrastructure
- Compromise or loss of critical data
- Erosion of company reputation

6.4.2 Vulnerability

Vulnerability is a measure of the likelihood of adversary success (L_{AS}) in causing the desired consequences (mathematical complement of protection system effectiveness). Vulnerability, or L_{AS}, is an estimate of the likelihood that the existing security countermeasures will be overcome by the attempted attack. This factor represents a qualitative surrogate for the quantitative conditional probability of success used in some government risk assessment models (U.S. Department of Justice, 2002). There are numerous subfactors involved in the analysis of L_{AS}, so this factor may be difficult to quantify. Alternatively, the TSVA team can assess the vulnerabilities and existing countermeasures to determine the defined adversary's ability to succeed.

6.4.3 Threat

Threat can be defined as any indication, circumstance, or event with the potential to cause loss of, or damage to, an asset at a specific location along the transportation route. It can also be defined as the intention and capability of an adversary to undertake actions detrimental to valued assets. Sources of threats may be categorized as:

- Foreign government
- Disgruntled employee or contractor
- Criminal
- Violent activist
- Terrorist (political, religious, environmental)

Each type of threat type may be categorized into three general groups—insiders, outsiders, and insiders working as colluders with others. Depending on the threat, the analysis team can identify the types of potential attacks and, if specified information is available (intelligence) on potential targets (chemicals, routes, and/or specific segments along a route), the likelihood of an attack and the specific countermeasures necessary to reduce undesired consequences. Often, this information may be too vague to be useful, but companies are encouraged to seek available information from other companies in the supply chain and from federal, state, and local law enforcement officials in analyzing threats. In the absence of specific threat information, the TSVA can still assume general capabilities and characteristics of typical hypothetical adversaries from the ones mentioned above.

Another component of threat is target attractiveness (A_T). Not all targets (chemicals, modes of transport, routes, or route segments) are of equal value to adversaries and this distinction is another factor that influences the likelihood of a chemical security event. A_T is an estimate of the real or perceived value of a target to an adversary based on factors such as:

- Potential for mass casualties/fatalities
- Extensive property damage
- Proximity to a national asset or landmark
- Possible disruption or damage to critical transportation infrastructure
- Possible disruption of national, regional, or local economy
- Ease of access to target
- Extent of media interest
- Company reputation and brand exposure
- Iconic or symbolic target

Particularly for terrorist acts, experience has shown that some types of assets are more likely to be targeted than others because they better accomplish the adversaries' objectives. During a TSVA, consideration may be given to a qualitative broad estimate of A_T for a specific chemical or route. Rather than attempt to calculate the actual likelihood that an adversary will attack a particular location along the transit route (this calculation is difficult due to a lack of reliable intelligence data), surrogate factors can be used. These surrogate factors enable the relative ranking of targets or locations along a transportation network as more or less attractive to adversaries, rather than using a likelihood of adversary attack (L_A) estimate, which is a factor used in some SVA models.

6.4.4 TSVA Methodology

There are several SVA techniques and methods available to the industry, all of which share common elements. Below is a list of some available SVA methodologies published by various governments, private groups, and trade/professional organizations. Some are merely chapters or sections of documents that address security or risk assessment/risk management in broader terms. Some are SVA publications by themselves. Some of these "methods" are complete, systematic analytical techniques and some are more checklist-type reviews.

- American Institute of Chemical Engineers, Center for Chemical Process Safety, *Guidelines For Analyzing And Managing The Security Vulnerabilities Of Fixed Chemical Sites, Second Edition*, 2002
- American Petroleum Institute/National Petroleum Refiners Association (API/NPRA), *Security Vulnerability Assessment Methodology for the Petroleum Industry*, 2004
- National Institute of Justice, *Chemical Facility Vulnerability Assessment Methodology*, July 2002 (Sandia VAM-CF)
- Synthetic Organic Chemical Manufacturers Association, Inc. (SOCMA), *Manual on Chemical Site Security Vulnerability Analysis Methodology and Model*, 2002
- U.S. Transportation Security Administration (TSA):
  - Transportation Risk Assessment and Vulnerability Evaluation Tool
  - Maritime Vulnerability Identification Self-Assessment Tool
  - Mass Transit Vulnerability Identification Self-Assessment Tool

The risk being analyzed in a TSVA is an expression of the likelihood that a defined threat will exploit a specific vulnerability of a particular target or combination of targets to cause a given set of consequences along a transportation route. Since historical data and trends (available for accidents) do not directly apply to intentional acts, a security risk analysis is usually performed using qualitative techniques. The expected outcome is a qualitative estimate of risk that can be used as a basis for determining which security issues may require additional protective countermeasures.

Vulnerability assessment is not a predictive science—it is an objective way of analyzing subjective issues. Given these realities and the evolving threat environment, the CCPS SVA method, with the specific changes made to address transportation issues, provides a transparent and flexible set of tools to evaluate a wide variety of chemicals, modes of transportation, routes, and to efficiently determine vulnerabilities, assess related risks, and evaluate countermeasures to minimize those risks.

The general TSVA steps are detailed in Figure 6.2, and include:

- *Step 1—Route Characterization*
  - Security prioritization process (Section 6.3).
  - Escalation of issues.
- *Step 2—Threat Assessment*
  - Identify and characterize adversary.
  - Target attractiveness determination.
  - Select target (segment of the route) for inclusion in analysis
- *Step 3—Vulnerability Analysis*
  - Define security scenarios.
  - Evaluate specific consequences.
  - Evaluate existing security measure effectiveness.
  - Identify and estimate degree of vulnerability.
- *Step 4—Risk Analysis*
  - Estimate likelihood of attack (combination of vulnerability, threat, and attractiveness) or conditional risk.
  - Estimate risk (risk index or risk matrix techniques).
  - Evaluate need for additional countermeasures (comparison to evaluation criteria).
- *Step 5—Countermeasure Analysis*
  - Identify and evaluate potential countermeasure options.
  - Prioritize potential security upgrades by cost, effectiveness, and other factors.

This book is not intended to be the singular reference for transportation vulnerability assessment, but rather provide the background for a transportation professional supporting such an analysis. Therefore, the SVA references presented above will provide more detail, with the understanding that most were developed from the perspective of chemical security for fixed chemical sites. The exceptions are the TSA tools and API/NPRA SVA methodology, which addresses both fixed-facility and transportation risks. Regardless, the specific security issues detailed in this chapter can be incorporated into any vulnerability assessment methodology selected. It is foreseeable that current fixed-facility methods will be expanded to include transportation issues, as well as new TSVA approaches developed.

Figure 6.2 Transportation SVA Methodology Steps

6.4.5 TSVA Resources

TSVAs depend upon a well-organized, documented discussion process and the availability of trained, highly experienced safety, security, and transportation personnel to yield reliable, replicable, and valid results. The process is a team-based approach that combines the participants' skills and knowledge to provide a complete security analysis of the transportation operation. Depending on the complexity of the transportation system, the team may include individuals with knowledge of physical and cyber security, process safety, logistics, operations, emergency response, management, and other disciplines as necessary. Additionally, the team should include representatives from all the key stakeholders, including carriers, other supply chain partners, customers, as necessary.

6.5 PRACTICAL TRANSPORTATION SECURITY ELEMENTS

A variety of ideas, recommendations, and practices put forward by government, transportation agencies, and industry members suggest security measures for consideration as part of a security plan and throughout the entire transportation network. The list provided below is not exhaustive, and some measures may not be appropriate for all companies or transportation activities. Each company and its supply chain partners will need to develop a security plan and security countermeasures based on their culture, baseline security measures, product, customers, and transportation issues.

6.5.1 General Security Elements and Security Plans

- Gather data about your own operations and those of other businesses with similar product lines and transportation patterns.
- Establish foundational security program elements.
- Increase security measures when there is a heightened alert (e.g., DHS Homeland Security Advisory System (HSAS), MTSA Maritime Security (MARSEC) Levels), a breach in security, or other suspicious activities.
- Develop a communications network to share best practices and lessons learned.
- Share information on security incidents to determine if there is a pattern of activities.
- Revise your security plans, as necessary, to take into account changing circumstances and new information.
- Conduct background checks during the hiring process and on current employees (specifically those with access to security-sensitive information) in accordance with regulatory requirements and your company's policies.
- Communicate regularly with employees about security, including current events and the need for employee vigilance.
- Pre-approve e-customers or entry of e-orders into IT systems by customer service representatives rather than allowing automated entries.
- Review process for accepting new customers, and change it, if necessary due to security issues.
- Review process to follow up with customers on a regular basis.
- Provide security training to all individuals commensurate with their security responsibilities. For the general employee population, this consists of security awareness training, including recognition and reporting of potential security risks. For others with specific security duties, this includes more detailed training on those responsibilities.

A security plan should include, at a minimum, personnel security, unauthorized access, in-transit security, and any additional regulatory requirements.

- Personnel Security
  - Perform background checks.
  - Encourage employees to report suspicious incidents or events.
  - Implement routine security inspections.

- Unauthorized Access
  - Establish partnerships with local law enforcement officials, emergency responders, and other public safety agencies with jurisdiction over your facility.
  - Request a review of your facility and security program by local law enforcement and fire safety officials.
  - Restrict the availability of information related to your facility and the materials you handle, and encourage authorities in possession of information about your facility to limit disclosure of that information to a need-to-know basis.
  - Check the adequacy of locks and other protective equipment.
  - Install additional lights, alarm systems, or surveillance cameras where needed.
  - Restrict access to your site.
  - Place limits on visitor access; require visitors to register and show photo identification and accompany visitors at all times.
  - Require employees to display identification cards or badges.
  - Conduct security spot checks of personnel and vehicles.
  - Upgrade security procedures for handling pick-ups and deliveries at your facilities. Verify all paperwork and require pick-ups and deliveries be handled only by appointment with known vendors. Require vendors to call before a delivery and provide the driver's name and vehicle number. Accept packages and deliveries only at approved gates.
  - Secure hazardous materials in locked buildings or fenced areas. Have a sign-out system for keys.
  - Secure valves, manways, and other fixtures on transportation equipment when not in use. Lock all vehicle and delivery trailer doors when not in use. Secure all rail, truck, and intermodal containers when stored at your location.
  - Use tamper-resistant or tamper-evident seals and locks on cargo compartment openings.
  - Periodically inventory the quantity of hazardous materials you have on site in order to recognize if a theft has occurred.
  - Keep records of security incidents. Review records to identify trends and potential vulnerabilities.
  - Report any suspicious incidents or individuals to your local law enforcement officials.
  - Use separate lanes for private cars and cargo vehicles
  - Use obliged routes for vehicles

- In-Transit Security
  - Use carrier safety ratings, assessments, safety surveys, or audits, and review security measures with carriers.
  - Verify carrier has an appropriate employee hiring and review process, including background checks, and an ongoing security training program.
  - Verify identity of the carrier and/or driver prior to loading a hazardous material.
  - Require driver photo identification (e.g., commercial driver's license, TWIC ID, FAST card) for comparison with information provided by the carrier.
  - Review with driver the name of the consignee and the destination for the material and confirm with your records before releasing shipments.
  - Identify preferred and alternative routing, including acceptable deviations.
  - Strive to minimize exposures to people and critical infrastructure, while expediting shipment to destination.
  - Minimize stops in transit. If stops are required, select locations with adequate lighting on well-traveled roads, and check vehicle after each stop to make sure no tampering has taken place. Consider stops only at appropriate locations (truck stops instead of public facilities).
  - Consider using two drivers or driver relays to minimize stops. Avoid layovers, particularly for high-hazard materials.
  - Work with rail carriers to identify roles and responsibilities for the security of rail cars stored temporarily on leased tracks. Sites receiving dangerous goods by rail should require rail cars to be delivered within a secured facility boundary, rather than dropped off in an unsecured location.
  - If materials must be stored during transportation, make sure they are stored in secure facilities.
  - Provide training to drivers to avoid highjacking or stolen cargo. Keep vehicles locked when parked and avoid casual conversations with strangers about cargoes and routes.
  - Consider whether a guard or escort for a specific shipment of hazardous material is appropriate.
  - Consider using advanced technology to track location or condition of shipments en route to their destinations.
  - Consider using radio-frequency identification (RFID) or other tracking methods for package (non-bulk) shipments of sensitive materials.
  - Consider installing tamper-proof seals on all valves and package or container openings.
  - Establish a communication system with transport vehicles and operators, including a crisis communication system with primary and

back-up means of communication among the shipper, carrier, and law enforcement and emergency response officials.

- Implement a system for a customer to alert the shipper if a hazardous materials shipment is not received when expected.
- When products are delivered, check the carrier's identification (driver and vehicle) with the shipping documents.
- Know your customers and their hazardous materials programs.
- Notify your local law enforcement officials of suspicious orders.
- Report any other suspicious incidents or individuals to your local law enforcement officials.

6.5.2 Additional Considerations

- Shipping and Receiving
  - Before shipping
    - Secure and safe storage in preparation for shipment release.
    - Work with customer and carrier to establish realistic arrival times/dates to expedite getting shipment to destination. Avoid scheduling that requires extended layovers, such as over a weekend.
    - Establish vehicle/driver screening/inspection process for entry to site; identify roles and responsibilities and methods/scope of inspection.
  - Before receiving
    - Establish realistic arrival times/dates to expedite getting shipment on site (i.e., avoid lines of trucks off site or tank cars staged outside plant).
    - Establish vehicle/driver screening/inspection process for entry to site.
    - Identify roles and responsibilities and methods/scope of inspection.
    - Establish a QA/QC process and/or Certificate of Analysis reviews.
    - Establish a plan to address non-conformances (e.g., tank car arrives with no seal or broken seal).
  - Security incidents
    - Establish the criteria for reporting security incidents.
    - Consider complementary (if not common) reporting and investigation procedures and systems for safety and security.

6.6 XYZ CHEMICAL EXAMPLE—SECURITY ANALYSIS

In Chapter 3, a corporate-level identification and prioritization of all worldwide hazardous material transportation activities was conducted by XYZ Chemical. The chemicals listed in Table 6.6 were identified as potentially high risk based on the possible impacts of a chemical release along the transit route.

As discussed in Chapter 3, the results of the initial company-level prioritization were based solely on the chemical hazards and identified facilities whose transportation operations would require a facility/operational level review. The focus of this example is on the security of the hazardous materials in and out of a single XYZ Chemical facility, and is the same Asian plant that was evaluated for safety issues in Chapters 4 and 5. In addition to the corporate directive to evaluate transportation safety risk at this facility, the site security manager was informed of the need to complete a security risk analysis of the hazardous materials in transit. To initiate the security review, the security manager and health and safety manager met to review the findings of the safety analysis. In addition to focusing on chemicals and hazards from the initial safety prioritization, the security prioritization process also considers:

- Modes and quantities transported
- Routes and movements
- Sensitive areas along the route

Table 6.6 XYZ Chemical Prioritization Process Results

Chemicals in Transit	Hazards	Potential Impacts	Risk Ranking
Chlorine	Hazard Division 2.3 (toxic gas)	Toxic exposure to people located along transportation routes	**High**
Pesticides	Hazard Class 6 (toxic substance)	Toxic exposure to people located along transportation routes	**High**
Ethylene Oxide	Hazard Division 2.3 (toxic gas) Hazard Division 2.1 (flammable gas)	Potential for toxic exposure, vapor cloud fire, fire, or VCE	**High**
Herbicides	Hazard Class 6 (toxic substance)	Toxic exposure to people located along transportation routes	*Medium*
Aqueous Acids	Hazard Class 8 (corrosive)	Potential environmental impact	*Medium*
1,3-Butadiene	Hazard Division 2.1 (flammable gas)	Potential VCE and fire hazard	*Medium*
Acrylonitrile	Hazard Class 3 (flammable liquid)	Potential VCE and fire hazard	*Medium*

While the identification of issues and escalation in the safety reviews were based primarily on loss-of-containment events, the prioritization of the security issues added the following issues or potential concerns:

- Theft, diversion, and misuse with the intent to cause harm along the route or at an alternate location
- Contamination with the intent to cause a reactive event while in transit, an event at the final destination, or other off-site impacts that could result from tampering with the material

Table 6.7 compares the issues that were screened out through the prioritization process, evaluated through benchmarking, and escalated for a more detailed risk analysis in both the safety and security prioritization processes.

Table 6.7 illustrates that the prioritization processes for safety and security screened out and escalated different issues. From a safety perspective, both the repackaged chlorine cylinders and the tank trucks of raw material butadiene and acrylonitrile were screened out. But, from a security perspective, repackaged chlorine (both cylinders and ton containers) poses a theft risk, and the flammable raw materials could be contaminated while in transit to the plant. While the railcar shipments of chlorine and ethylene oxide and pesticide truck shipments were not screened out initially, a security benchmarking approach was selected for these operations as compared to escalating them to a full TSVA. As part of this benchmarking activity, the security regulations in this foreign market will be reviewed and the security countermeasures employed in the United States and Europe. Since these regions have documented security regulations, these will set the baseline for the developing countries where XZY Chemical has operations.

Table 6.7 Comparison of Safety and Security Prioritization

Prioritization Result	Safety Analysis	Security Analysis
Screened out	• Herbicide product shipments • Waste stream of 32% HCl • Butadiene and acrylonitrile truck receipts • Repackaged chlorine cylinders	• Herbicide product shipments • Waste stream of 32% HCl
Benchmarking	• Chlorine rail safety practices	• Rail security practices (chlorine and ethylene oxide) • Bulk and non-bulk pesticide truck shipments
Escalated for more detailed review	• Repackaged chlorine in ton containers • Bulk and non-bulk pesticide shipments • Ethylene oxide rail car deliveries	• Repackaged chlorine • Butadiene and acrylonitrile truck receipts

The transportation operations escalated for a more detailed TSVA at this facility are illustrated in Figure 6.3.

Figure 6.3 Operations Escalated from Security Prioritization Process

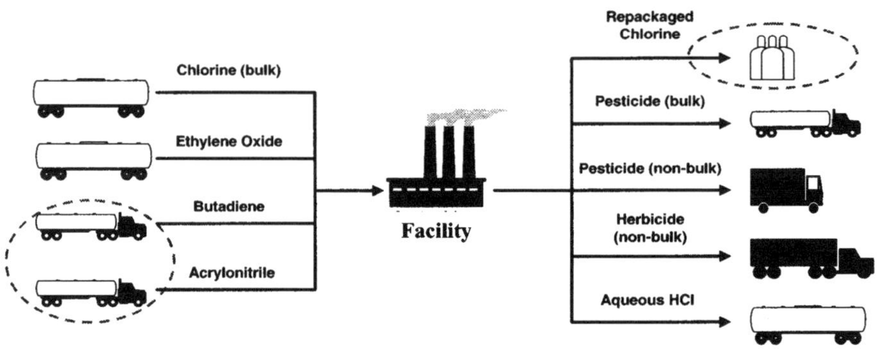

Following the security prioritization of the plant's transportation operations, the following three chemicals and modes of transportation were escalated for a more detailed TSVA:

- Repackaged chlorine in 150-pound cylinders and ton containers
- Tank truck deliveries of butadiene
- Tank truck deliveries of acrylonitrile

To complete the TSVA, a corporate risk management representative came to the facility to lead a local team through the analysis methodology. The team received background information on the risk matrix and the vulnerability methodology that would be followed. The interdisciplinary team members represented the following departments:

- Security
- EHS
- Logistics
- Information technology (IT)
- Purchasing
- Butadiene and acrylonitrile production
- Butadiene and acrylonitrile carriers
- Emergency response

As directed by corporate, and consistent for all XYZ Chemical facilities conducting TSVAs, the following were defined for the security analysis:

- *Scope of Analysis*—limited to chemicals and modes escalated from the security prioritization process

- *Threats*—no specific threats will be analyzed (e.g., likelihood of attack not considered); security analysis will be a conditional risk analysis focused on consequences and security vulnerabilities
- *Scenarios*—all applicable security scenarios will be evaluated
- *Analysis Guidelines*—for consistency with other facilities, corporate provided the consequence, vulnerability (the likelihood of attack success), evaluation criteria, and risk matrix

The risk matrix for this facilitated analysis is illustrated in Figure 6.4 and is consistent with the safety risk matrix from Chapter 4. The consequence, likelihood (of success or vulnerability), and evaluation criteria definitions are detailed in Tables 6.8–6.10, respectively.

Figure 6.4 XYZ Chemical Risk Matrix

	1	2	3	4
4	*IV*	*II*	*I*	*I*
3	*IV*	*III*	*II*	*I*
2	*IV*	*IV*	*III*	*II*
1	*IV*	*IV*	*IV*	*III*

Consequence Category

→

Increasing Severity

Table 6.8 XYZ Chemical Consequence Categories

Consequence Level	Description
1. Minor	• Safety: No or limited minor injuries • Environmental: No or small spill response • Property: Damage < US$500,000
2. Low	• Safety: Potential multiple minor injuries or limited serious injuries • Environmental: Large spill response • Property: US$500,000–US$1 million
3. High	• Safety: Potential multiple serious injuries or limited fatalities • Environmental: Short-term remediation • Property: US$1 million–US$10 million
4. Very High	• Safety: Potential multiple fatalities • Environmental: Long-term remediation • Property: Damage > US$10 million

Table 6.9 XYZ Chemical Likelihood Categories

Likelihood Level	Description
1. Extremely Unlikely (or remote)	Multiple layers of effective security countermeasures
2. Very Unlikely	Single layer of effective security countermeasures
3. Unlikely	Single, but effective, security countermeasure
4. Likely	Limited or no effective security countermeasures

Table 6.10 XYZ Chemical Evaluation Criteria

Number	Category	Description
I	Action Required – High Priority	Risk mitigation options must be recommended
II	Action Required	Risk mitigations should be considered
III	Tolerable with Controls	Procedures and controls verified
IV	Tolerable as is	No mitigation required

The results of the TSVA are detailed in Table 6.11.

Table 6.11 XYZ Chemical TSVA Results

Chemical	Security Concern	Location	Consequence	Likelihood	Risk Level
Chlorine (cylinders)	Theft	On site En route Ordering (cyber)	3	2 3 3	III II II
Chlorine (tons)	Theft	On site En route Ordering (cyber)	4	1 2 3	III II I
Butadiene	Contamination	Production site En route	3	3 3	II II
Acrylonitrile	Contamination	Production site En route	2	3 3	III III

From these results, the following was determined:

- Repackaged chlorine has three security issues surrounding theft: 1) on site at the facility, 2) in transit, and 3) through the ordering system (or through cyber means).
  - Consequences:
    - Theft of chlorine cylinders and ton containers could result in an attack at another location using chlorine as a weapon. While both quantities could result in fatalities, ton containers were ranked as a higher consequence level.
  - Vulnerabilities:
    - On site: Cylinders are monitored in the repackaging area and warehouse, but are not secured until loaded for delivery.
    - In transit: Drivers are permitted to stop along the route, and there is no monitoring of the trucks while in transit.
    - Ordering system: No product stewardship program or formal customer verification process.
  - Risk results:
    - The on-site security risk is identified as tolerable. Risk ranking of III.
    - The in-transit shipments of chlorine are identified as a higher risk with action required. Risk ranking of II.
    - The ordering system for chlorine is identified as a higher risk with action required. Risk ranking of II for cylinders and I for containers.
- Contamination is an issue both at the production facilities for butadiene and acrylonitrile and in transit:

- Consequences: Contamination of these materials could lead to a release and secondary damage to property from a localized fire or BLEVE along the transit route.
- Vulnerabilities: Monitoring of the loading process at the production site, but no means of verifying tampering (e.g., no seals and no ability to determine if tampering has occurred in transit).
- Risk results:
    - Butadiene contamination (both at the production site and in transit) is identified as a higher risk with action required. Risk ranking of II.
    - Acrylonitrile contamination (both at the production site and in transit) is identified as a tolerable risk. Risk ranking of III.

Based on these results, the team recommended the following security countermeasures for consideration:

- Chlorine:
    - Consider not permitting stops during the delivery of repackaged chlorine or require drivers to radio in when they have stopped, report their location, and verify the security of the shipment.
    - Consider developing a formal product stewardship program for all current and future chlorine customers.
    - Consider a follow-up cyber vulnerability assessment to determine if the system is vulnerable to attack and the creation of false orders.
- Butadiene:
    - Consider requiring supplier to add tamper-resistant seals to all tank trucks, and require the facility to verify the seal before accepting a tank truck for on-site delivery.

The selection of the appropriate security countermeasures will be included as part of the conclusion of the example in Chapter 7.

REFERENCES

American Chemistry Council, *Implementation Resource Guide for Responsible Care® Security Code of Management Practices: Value Chain Activities*, September 2002

American Chemistry Council, The Chlorine Institute, National Association of Chemical Distributors, *Transportation Security Guidelines for the U.S. Chemical Industry*, 2001

American Petroleum Institute, National Petrochemical & Refiners Association, *Security Vulnerability Assessment for the Petroleum and Petrochemical Industries, Second Edition*, August 2004

Center for Chemical Process Safety, *Guidelines for Analyzing and Managing the Security Vulnerabilities of Fixed Chemical Sites*, American Institute of Chemical Engineers, New York, 2003

United Nations Economic Commission for Europe, *Recommendations on the Transport of Dangerous Goods, Model Regulation Volume I, Fifteenth Revised Edition*, United Nations, New York and Geneva, 2007

U.S. Department of Homeland Security, *Chemical Facility Anti-Terrorism Standards*, Interim Final Rule, 6 C.F.R. Part 27, April 2007

U.S. Department of Justice, *Chemical Facility Vulnerability Assessment Methodology*, NIJ Special Report, Office of Justice Programs, National Institute of Justice, July 2002

U.S. Department of Transportation, *Hazardous Materials Transportation Enhanced Security Requirements*, Research and Special Programs Administration, Washington D.C., DHM50-0030-0903, 2007

7
RISK REDUCTION STRATEGIES

In the previous chapters, analysis techniques were presented for identifying and evaluating the safety and security risks associated with the transportation of hazardous materials over a route(s) and across different modes of transit. Any of these techniques can be used to determine where additional risk reduction options might be warranted and to identify and evaluate potential options. Additional risk reduction opportunities may include identified inconsistent operating practices, changes/advances in industry guidelines, or new regulatory requirements.

The value of a systematic approach to risk management—as has been presented in this book—is to help ensure that risk reduction resources are applied appropriately to address the highest priority risks. Although it may be tempting to bypass the prioritization and risk analysis steps, and to arbitrarily implement some of the commonly practiced risk reduction strategies presented in this chapter, risk analysis provides valuable information and insights that can be incorporated into a comprehensive decision-making process. Whether the need to reduce risks has been warranted through analysis or through management or regulatory dictate, an action plan to identify, evaluate, compare, and select appropriate risk reduction options should be prepared.

This chapter provides guidance on developing risk reduction strategies and highlights the factors that influence the different types of safety and security measures that can be selected and ultimately implemented. Additionally, given the complexity of a transportation network and acceptance needed among all stakeholders, risk reduction opportunities should be evaluated explicitly in the context of the benefits, comparison to other options, and the costs.

7.1 RISK REDUCTION INITIATIVES

As a natural result of applying prioritization and qualitative risk analysis techniques, risk reduction options are identified. This is accomplished by reviewing the general transportation program elements and benchmarking activities where practices are compared and contrasted. Simply reviewing where gaps exist may quickly and effectively identify actions that should be considered to reduce risk.

Likewise, monitoring the baseline transportation safety and security management systems for changes in operating conditions may identify the need for additional risk mitigation efforts. Risk management needs can be identified outside of the risk analysis techniques presented in Chapters 4, 5, and 6 by tracking the following trends:

- An increase in carrier accident rates
- Problems identified during inspections
- Violations of management practices
- Changes in the number of movements
- New customers
- New chemical movements
- Changes in routes or modes
- Specific threat information
- Changing population trends/demographics along route

This list is not comprehensive and can take many forms depending on the specific design of an organization's management system and its key control parameters.

As these parameters are monitored and changes in risk identified, critical issues can be escalated for more detailed review. Once several risk reduction strategies are identified, the same types of risk evaluation criteria (e.g., risk index, risk matrix, or other quantitative measures) described earlier in this book can be used to assess the relative benefits of each proposed risk mitigation option. Risk reduction can thus be defined as the process of evaluating and identifying options available to reduce risk, that achieve the desired level of risk reduction, and can be justified on a cost-benefit basis.

7.1.1 Documentation and Communication of Risk Reduction Strategies

Documentation and communication are the keys to a successful risk management program and the implementation of a successful mitigation strategy. This is true regardless of the specific risk analysis techniques and the reduction options that are being considered and finally implemented. Written documentation for all risk management activities will justify, support, and provide a transparent record for each of the following risk management activities:

- Risk management process
- Risk analysis protocol
- Scope and objectives
- Risk analysis assumptions and uncertainties
- Risk evaluation criteria
- Stakeholder involvement
- Risk results

- Risk reduction options for consideration
- Risk reduction approval and final actions
- Date risk reduction options initiated and completed
- Timing for periodic risk review and revalidation

Each step in the transportation risk management process (e.g., primary management systems, prioritization, risk analysis, risk reduction options, and process sustainability) requires valid and accurate input to ensure the results can be used as part of management's decision process. Therefore, risk activities should include other internal business units and transportation stakeholders in the process. Internal communication benefits the risk management process through:

- Verification and alignment with company objectives
- Knowledge of the organizational structure to identify critical functions and people
- Collection of current safety and security procedures and risk assessment protocols
- Comparison to the appropriate risk evaluation criteria
- Identification of employee and business unit responsibilities

External communication with other stakeholders involves carriers, suppliers, and customers working together to evaluate the transportation network from an overall perspective. This communication process may include:

- Identification of key stakeholders (e.g., vendors, logistic service providers (LSPs), emergency response providers, customers)
- Communication of objectives to stakeholders
- Integration of desired procedures into day-to-day operations
- Development of dialogue with partners

7.1.2 Risk Reduction Opportunities

Opportunities for reducing risk can be identified at any point in the transportation risk management process. A more detailed transportation risk analysis, including full quantification, is generally not required to identify and recommend operational improvements. As discussed, options for improvement may be recommended as part of a review of the primary management systems (e.g., regulatory, training, protocols and procedures, equipment integrity, monitoring, incident preparation), or during the prioritization and qualitative screening activities. As illustrated in the Figure 7.1, when risk reduction (or mitigation) options are identified, they need to be incorporated into the risk analysis process. This will help to ensure that the risk reduction options being considered are evaluated consistently, as compared with the current (or baseline level) operational level of risk, and that the desired level of risk reduction can be achieved. Specific risk reduction options are presented in Section 7.2.

Figure 7.1 Risk Reduction Identification and Evaluation

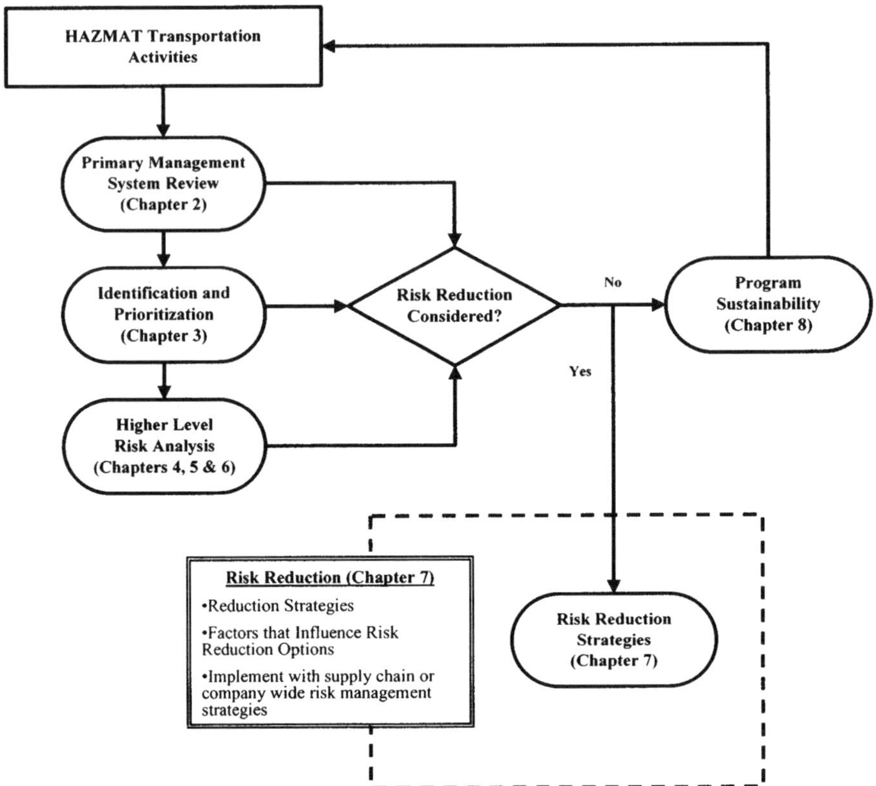

7.1.3 Balancing Safety and Security

Risk reduction options for the safe transportation of hazardous materials can include a decrease in the potential consequence, the likelihood of an in-transit incident (accidental or non-accidental causes), or a risk reduction option that impacts both consequence and likelihood. Options to reduce transportation safety risk can be grouped into the following four major categories:

- *Detect*—a safety strategy to identify potential issues before an incident occurs. This may include periodic reviews or audits of the operation and its carriers as well as ongoing risk analysis activities to identify new opportunities to improve safety and reduce risks. Additionally under this category, detection can include techniques such as GPS tracking, which, while not preventing or mitigating a release, allows faster identification of asset location.
- *Prevent*—a safety strategy focused on eliminating the likelihood of an accident and/or a resultant level of consequence. Prevention may include producing the chemical at the customer site, thus eliminating the

transportation all together, or shipping another material/intermediate with significantly lower hazards.

- *Mitigate*—a safety strategy focused on managing the risk through decreasing the potential likelihood of an in-transit release and/or decreasing the potential consequences. Mitigation may take the form of carrier or route selections to minimize the chance of an accident and changes in the mode or container design to decrease the likelihood or potential size of a release.
- *Respond*—a safety strategy focused on responding to an in-transit accident in an attempt to mitigate the potential release extent and the resulting impacts to the surrounding community, infrastructure, or environment. This category also includes technologies that detect and report when a release has occurred, facilitating quicker and more effective emergency response.

In addition to mitigating and preventing the potential consequences, security strategies often employ four basic objectives to help minimize the attractiveness, vulnerability, and threat:

- *Deter*—a security strategy to prevent or discourage the occurrence of a breach of security by means of fear of being detected or doubt in the ability to achieve a successful attack
- *Detect*—a security strategy to identify an adversary attempting to commit a malicious act or other criminal activity in order to provide real-time observations, interception, and post-incident analysis of the activities
- *Delay*—a security strategy to provide various barriers that slow the process of an adversary gaining access to the shipment to prevent an attack or theft, or to assist in apprehension and prevention of theft
- *Respond*—a security strategy to neutralize the adversary, or to evacuate, shelter in place, call local authorities, control a release, or take other actions

Appropriate strategies for managing security risk can vary widely depending on the individual circumstances of the chemical, transportation mode, route, carrier, threats, etc. The specific situations must be evaluated individually and managed based on the collective judgment of the supply chain partners, always balancing safety and security.

Risk reduction options must be evaluated to understand how the additional security measures will affect the safety of the operation, as well as how additional safety measures could impact security. For example, rerouting a chemical shipment around a major metropolitan area for security purposes may result in an important change in the route. This change could result in a significant increase in the route length, with an overall increase in the likelihood of a transportation accident. This change in the safety risk profile needs to be compared to the benefits of the change based on security. This comparison will help to ensure that the proper tradeoff is selected and that risk is not unknowingly transferred from one issue to another.

7.2 FACTORS INFLUENCING RISK REDUCTION OPTIONS

A number of factors should be considered when recommending and selecting risk reduction options. These may include:

- Regulatory requirements
- Industry practices
- Societal expectations
- Feasibility, effectiveness, and reliability
- Availability of the mitigation option
- Skill and ingenuity of workforce
- Automated technology vs. manual application
- Cost-benefit

Both pre-shipment and emergency response planning and preparedness activities are critical to managing the risks associated with hazardous materials transportation. Additionally, understanding the entire value chain is critical to addressing the risks at various stages of manufacture, handling, transportation, use, disposal, etc. Therefore, risk management of these hazardous materials is much larger than just a transportation activity, and the implementation of risk reduction options may have impacts beyond the transportation activity being reviewed. Therefore, it is critical that changes in safety and security be communicated and understood across the business and supply chain. This will ensure that all requirements are being implemented and that unknown and undesirable tradeoffs have been identified and resolved.

7.2.1 Overall Supply Chain Risk Reduction Options

Opportunities to reduce risk across the overall supply chain may include:

- Shipment (inbound and outbound transport)
- Unloading and loading
- Product stewardship
- Business plan
  - Intermediate handling
  - Shifting production either to suppliers, in-house operations, or to customers to minimize the transport of the more hazardous materials

Table 7.1 lists some options to consider when evaluating these types of supply chain activities. This list is not exhaustive, but is intended to provide the user with the types of issues, options, benefits, and additional considerations when evaluating risk reduction.

Table 7.1 Risk Reduction Options—Supply Chain Activities

Issues	Risk Reduction Category	Risk Reduction Options	Benefits and Additional Considerations
Shipment	Mitigate	• Establish depots near customers • Bartering • Product trading	Reduction in the likelihood of accidents through better controlled and shorter loaded distances or use of the transportation infrastructure during off-peak times
	Mitigate	Optimize arrival times or staging of hazardous materials	Changes in sensitive receptors along the route (even if the distance is shorter)
	Mitigate	• Change mode • Change container type • Change supplier/carrier • Convert to company-owned/dedicated personnel and equipment	• Reduction in the likelihood of accidents • Specialized containers could also impact the probability of release, given an accident, and the spill size distribution
	Detect	Invite suppliers to provide product stewardship services for inbound hazardous materials	If new carrier or intermediate storage partners are considered, need to review their safety practices and performance
Loading and Unloading — Handling Requirements	Mitigate	Special procedures, equipment for loading and unloading	Potential reduction in the likelihood of accidents during loading and unloading
	Detect	Include loading and unloading as part of Process Hazards Analysis (PHA) activities	Adding these activities to the PHA enables risks to be identified and reduction measures to be implemented where necessary
Product Stewardship	Detect	Review and periodically audit customers to incorporate transportation safety and security elements into the product stewardship program	Ensure safe and secure handling of chemicals once custody transfer has occurred
Business Plan	Prevent	Assess possibility of transferring production either upstream or downstream to eliminate the transportation of higher hazard materials	May eliminate or significantly reduce risks associated with transportation

7.2.2 Pre-shipment Risk Reduction Options

Pre-shipment risk reductions are actions that can be taken when setting up the supply chain and the specifics of the operation, and may include:

- Defining the properties of the hazardous materials that can be shipped
- Identifying and specifying shipping points and receiving locations
- Defining acceptable modes of transportation for each hazardous material or classification of material
- Defining packaging and equipment requirements
- Specifying handling requirements
- Developing a list of approved carrier and logistical service providers

Risk reduction recommendations from the analysis of pre-shipment options may be the result of any or all of the parameters of the risk equations. As described in Chapter 3, risk is a function of the consequence and likelihood of the specific transportation scenario(s) under evaluation:

$$\text{Risk} = f(\text{scenario(s), consequence, likelihood})$$

Therefore, risk reduction options that have the potential to influence the consequence, likelihood, or both should be considered. Table 7.2 lists some commonly applied risk reduction options for consequence reduction. In addition to consequence risk reduction options, Table 7.3 lists commonly applied options for likelihood reduction. As can be seen, there is some overlap between these tables because some risk reduction options may impact both the consequence and the likelihood. These lists are not exhaustive, but provide an example of several pre-shipment options, as well as additional considerations when evaluating these types of risk reduction measures.

Table 7.2 Risk Reduction Options—Consequence

Options for Risk Reduction	Risk Reduction Category	Consequence Risk Reduction Option	Additional Considerations
Hazardous Material Properties	Prevent	Identify hazardous materials that will be prohibited from transit	Potential impacts to critical customers or supply chain disruption if material is no longer provided
	Mitigate	Modifying material properties or shipping intermediates to reduce potential consequences or releases	Intermediates may be less hazardous, but could require increased shipments or other factors that could increase the risk as compared to the original product
	Prevent/ Mitigate	Identify less hazardous substitutes	Less hazardous substitutes may have cost implications that do not justify the risk reduction benefits
	Prevent	Eliminate transportation activity by producing materials at customer or supplier facility	Cost-benefit, operation experience, change in facility risk profile
	Prevent/ Mitigate	Optimize shipment conditions to minimize potential consequences of release (e.g., temperature, pressure, phase, concentration)	Changes in production process and transportation carrier requirements
Shipment Points and Receiving Locations	Mitigate	Selection of routes that avoid/minimize population centers (metro areas)	Supply chain optimization could result in transit closer to sensitive population
	Detect	Develop product specific shipping specifications	If new carrier or intermediate storage partners are considered, need to review their safety practices and performance
Transportation Modes	Mitigate	Reducing quantity shipped	Reducing the shipping quantity could result in more deliveries and thus a higher potential for accidents and greater overall risk
	Mitigate/ Deter	Consider shipping the higher hazard materials via more secure mode (perhaps rail vs. truck)	Infrastructure and equipment may not be available to easily switch modes
	Mitigate	• Consider simplifying transportation process • Eliminate where possible transloading, off-loads (truck chassis to rail, etc.)	Requires optimization of supply chain activities

continued

Options for Risk Reduction	Risk Reduction Category	Consequence Risk Reduction Option	Additional Considerations
Packaging and Equipment	Mitigate/ Detect	Special safety features or controls	Customer needs/requirements
	Mitigate/ Prevent	Avoid shipment of hazardous materials with other non-hazardous public shipments (e.g., UPS, FedEx, DHL)	Mode of transport limitations
	Prevent	Avoid shipment of chemically reactive materials in the same shipment	Mode of transport limitations
Handling Requirements	Mitigate	In-transit analysis and specification of procedures for safety security	• Potential impact to on- site risks • Tradeoffs in routing considerations, specifically delivery times, quality of infrastructure, and time-sensitive materials

Table 7.3 Risk Reduction Options—Likelihood

Options for Risk Reduction	Risk Reduction Category	Likelihood Risk Reduction Option	Additional Considerations
Shipment Points and Receiving Locations	Mitigate	Optimize supply chain through a reduction in miles, reduce intermediate handling, or use routes/carriers with lower accident rates	• Supply chain optimization could result in transit closer to sensitive population • If new carrier or intermediate storage partners are considered, need to review their safety practices and performance
	Mitigate	Develop product-specific shipping specifications	
	Detect/ Deter	Establish trip plans for carriers, including acceptable locations for stops	
Transportation Modes	Mitigate	Reduce number of shipments	Tradeoffs need to be evaluated as reducing the number of shipments could result in larger shipment quantities, thus higher potential consequences if an accident were to occur
	Mitigate	Reduce the number of transfer points in supply chain	

continued

Options for Risk Reduction	Risk Reduction Category	Likelihood Risk Reduction Option	Additional Considerations
Packaging and Equipment	Mitigate	Design equipment above regulatory minimum when appropriate	• Special permits or approvals may be needed for more robust equipment design • Mode of transport may have unique limitations for changes in equipment design and ability for certain types of inspections
	Mitigate/ Detect	Use special safety features or controls	
	Mitigate	Increase equipment inspection intervals	
Carriers and Logistical Service Providers	Detect	Establish a robust carrier & terminal assessment selection program and ongoing audit program	• Available carriers cannot or are not willing to meet requirements • Need to communicate list of approved carriers and logistics service providers to ensure low-cost providers are not utilized without understanding the potential impacts to the operations risk profile
	Mitigate/ Detect	Document approved carriers (by product, region, etc.)	
	Mitigate/ Detect	Document approved logistical service providers	
	Detect	Evaluate customer pickups for safety and security practices	
	Mitigate	Use of dedicated carriers for certain cargos	
	Mitigate/ Deter	Requiring team drivers vs. layovers	
	Mitigate	Carrier experience requirements for hazardous materials carriers	
	Mitigate/ Prevent	• Selection of carriers/routes with lower observable accident rates • Work with carriers on improving accident rates	
	Mitigate	Set minimum requirements for driver training	
	Detect	Implement a carrier assessment program with periodic reviews/audits	
	Detect/ Deter	Track shipments	
Handling Requirements	Mitigate	Selection of routes that maximize the use of divided highways	Trade-offs in routing considerations, specifically delivery times, quality of infrastructure, and time-sensitive materials, all of which could impact the likelihood of transportation accidents
	Mitigate	Speed requirements based on mode and route	

7.2.3 Emergency Response and Post-Incident Risk Reduction Opportunities

Table 7.4 lists ways to learn from incidents and develop recommendations and systems to further reduce risk by helping to lower the chance of similar incidents in the future. Table 7.5 lists emergency response risk reduction opportunities. Although not exhaustive, this list provides several options to be considered and the thought process for documenting and comparing risk reduction options. As with all risk reduction options, these should be considered as part of the overall risk management process, and should not be selected arbitrarily or as a reactionary solution to a recent accident.

Table 7.4 Post-Incident Learning Opportunities

Learning Opportunities	Key Considerations
Reporting Processes	• Definition of an incident and spill (company vs. regulatory requirements) • Data/information to be reported (company vs. regulatory requirements) • Incident notification and initiation of emergency response • Communications/incident command
Investigation Processes	• Formal investigations (root cause analysis as warranted) • Communication of causes and corrective actions • Follow-up actions for verification that corrective actions were implemented
Incident Tracking and Analysis	• Database of incident information • Periodic review of data • Identification of trends in data • Share lessons learned internally and, as appropriate, externally • Proactively involve all parties to prevent future occurrences

Table 7.5 Emergency Response Risk Reduction Opportunities

Options for Risk Reduction	Key Considerations
Company Plan	• Clarification of roles and responsibilities • Communications—internally and externally • Actions—levels of response • Remote advisory, on-scene advisory, actual mitigation, etc.
Material Specific Plans	• Specific procedures • Equipment, personnel protection needs, etc. • Mutual aid agreements
In-House Response	• Level of training • Type of training (product, transport, equipment specific, etc.) • Periodic qualification • Equipment needs • Location and coverage area
Contract Response	• Assessment and selection • Capabilities, equipment, location, etc. • Periodic audit/review • Type and level of training (product, transport, equipment specific, etc.) • Media • Community evacuation
Drills/ Exercises	• Tabletop • Full-scale simulation • Internal and external safety and security responders
Community Involvement	• Engage emergency responders
International Considerations	• Determine and engage all appropriate agencies across various countries of transit • Determine capabilities of various agencies to develop coordinated response plan • Incorporate into detection, communication, and response plan

7.3 SELECTION OF RISK REDUCTION OPTIONS

The goal of this book is not to devise a specific set of risk reduction options, but, rather, to encourage the application of risk analysis techniques to identify where risk reduction is warranted, and to develop recommendations that can be used in the development of a customized risk reduction strategy. The suite of risk reduction options developed as part of the risk analysis activities should be thoroughly evaluated so that the appropriate mix of options is selected, one that balances the risk reduction goals and proper allocation of resources.

Once a set of recommendations has been developed, the options must be analyzed to determine the benefits, or essentially the level of risk reduction. The risk analysis method utilized to assess the baseline operational risks should be the same method used to analyze each of the potential risk mitigation options. Therefore, if a semi-quantitative technique was used to assess the baseline risk, then the same assumptions, techniques, and risk evaluation criteria (risk matrix, risk index, etc.) should be used to evaluate the effectiveness of the risk reduction options.

An example of this process is illustrated in Figure 7.2. In this example, an operation was evaluated with a base case risk level of I, or a "higher risk" that required consideration of risk reduction options. Three risk mitigation options were developed and evaluated. As can be seen, one option reduced the consequence (Option 1), one option reduced the likelihood (Option 2), and one option reduced both consequence and likelihood (Option 3). Option 3 resulted in the greatest risk reduction benefit. While the options that reduce only consequence or likelihood did reduce risk (movement across or down the matrix), the resulting level of risk, as defined on the risk matrix, did not change from risk level I. Therefore, purely on a benefit basis, Option 3 provided the greatest risk reduction.

Figure 7.2 Semi-Quantitative Comparison of Risk Mitigation Options

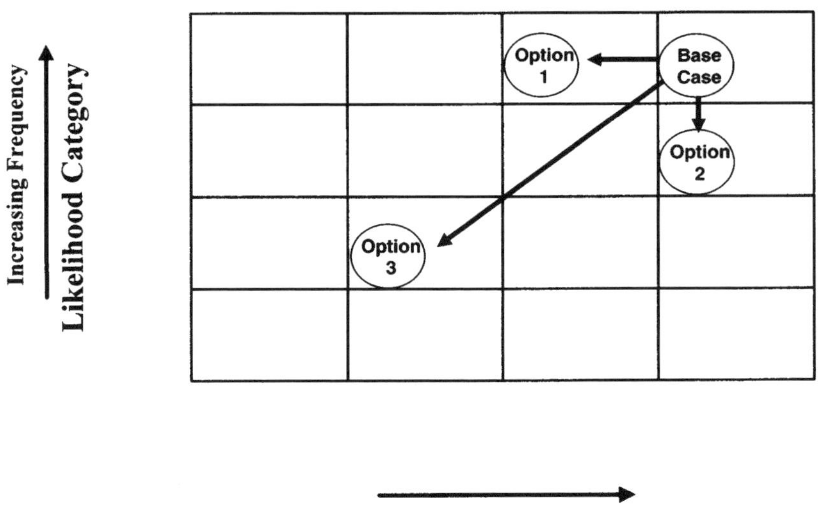

This information would next be input into the decision-making process, with the final set of risk mitigation options selected on a cost-benefit basis. It is possible that Options 1 and 2 are relatively inexpensive, required by regulation, or recommended by an industry standard. In these circumstances the low level of risk reduction benefit may not weigh as heavily in the decision process. Additionally if Option 3 is very costly, then both the benefits and the costs will be incorporated into the final decision.

Figure 7.3 shows another example that compares a base case level of operational risk with two risk reduction options using quantitative risk analysis techniques. This F-N curve from Chapter 5 illustrates that both options reduce the likelihood and the potential maximum consequence(s). However, Option 2 results in the greatest risk reduction benefits of lower maximum impacts and lower frequencies. Since both options result in risk reduction, as compared to the baseline, an additional cost-benefit analysis would be required to determine which (if either) of these two options should be considered for implementation.

Figure 7.3 Quantitative Comparison of Risk Mitigation Option

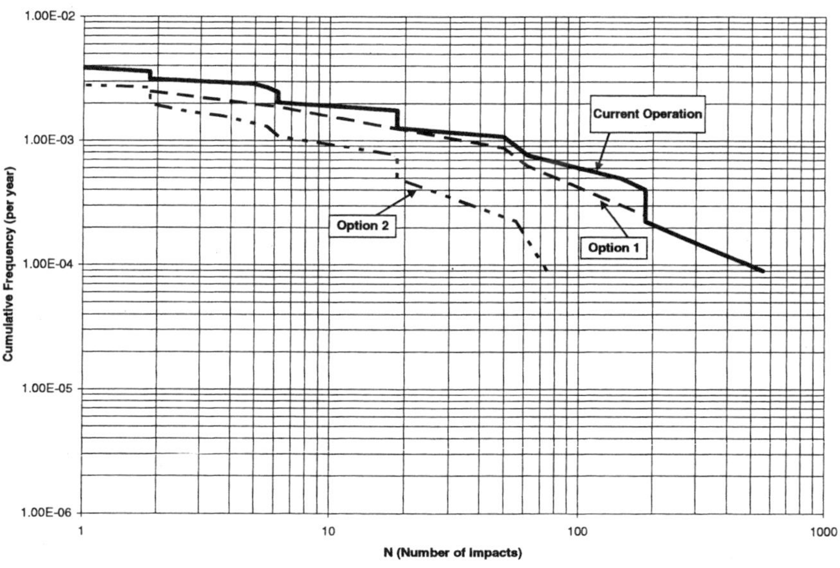

In addition to the risk reduction benefits, the costs of risk mitigation options need to be evaluated. Due to the uncertainties associated with semi-quantitative and quantitative risk analysis results, a relative risk comparison, as compared to absolute measures of risk and benefits, is recommended. To conduct this type of relative comparison, incremental risk analysis can be used to evaluate the cost effectiveness of risk mitigation options, or determine the optimal combination of risk mitigation options. Figure 7.4 illustrates example results of this type of analysis, and uses the options from the F-N curve in Figure 7.3 as the basis for comparison.

Figure 7.4 Incremental Risk Analysis

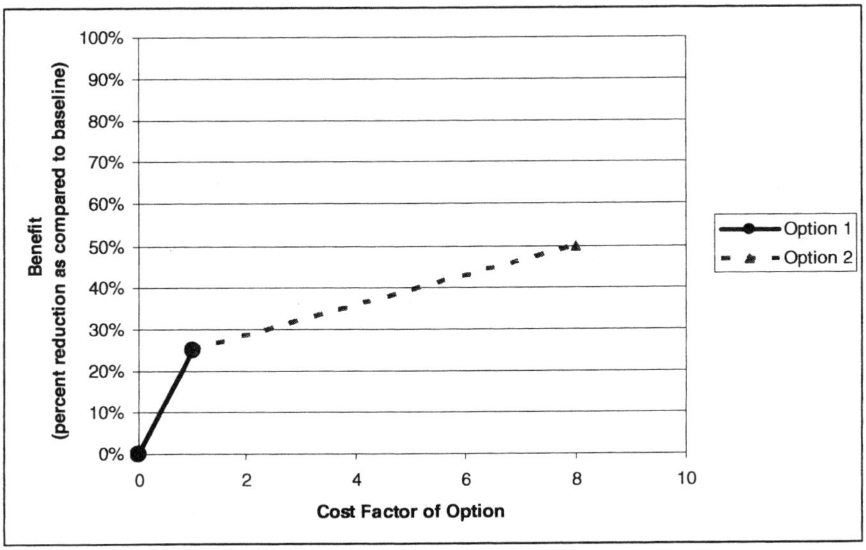

Figure 7.4 shows that risk reduction Option 1 results in a benefit of 25% reduction in risk, as compared to the baseline risk level. The level of risk calculated for the baseline operation and the options is the sum of the F-N pairs used to create Figure 7.5. The combination of Option 1 and Option 2 increases the benefit by an additional 25%, to a total of 50% reduction from the baseline, but at a cost eight times higher than Option 1 alone.. This type of information not only clarifies the risk mitigation associated with each design option, but also illustrates the comparison on a cost -benefit basis. Using the results from the F-N curve in Figure 7.3 alone may have influenced the decision that Option 1 should be selected over Option 2. However, considering both the benefits and costs enables a comparison that will result in better informed risk management decisions.

7.4 XYZ CHEMICAL EXAMPLE— RISK REDUCTION STRATEGIES

XYZ Chemical, the example company, and one of its facilities in the Asian market have been progressing through the transportation risk management (TRM) framework outlined in Chapter 1. As part of the review of the primary management systems, prioritization and identification of higher-risk operations, and safety and security risk analyses of escalated issues, several options to reduce risk for XYZ Chemical as a whole, and, specifically for the Asian facility, have been identified and documented. A summary of the risk mitigation options recommended at each stage of the risk management process is presented in Table 7.6.

Table 7.6 XYZ Chemical Risk Reduction Options

TRM Framework	XYZ Chemical Level of Benefit	Risk Management Issues	Risk Reduction Recommendation
Primary Management Systems (Chapter 2)	Corporate / Management Level	• No formalized global risk management process • Not all countries of operation have formalized safety and security programs	• Development of a corporate-wide transportation risk management system • Implementation of corporate-wide primary management system requirements for all counties of operation
Prioritization and Identification (Chapter 3)	Corporate / Management Level	• Hazards of the worldwide transportation operations are not clearly understood • Risk management priorities have not been identified	• Corporate-level prioritization of all transportation hazards • Identification of low-risk materials that can be limited to a primary management system review • Identification chemicals, modes of transportation, and specific facility operations escalated for more detailed risk analysis
Qualitative Risk Analysis (Chapter 4)	Facility Level	Asian facility transports chemicals escalated for detailed risk analysis	Based on benchmarking, implement bulk chlorine U.S. and European practice as a corporate standard
Semi-Quantitative Risk Analysis (Chapter 4)	Facility Level	Qualitative risk analysis resulted in the escalation of four of the nine chemical/ mode transportation operations from the Asian facility	• Increased detail resulted in screening out two of the chemical movements • Bulk chlorine and pesticide recommendations developed • Rerouting • Repackaging
Quantitative Risk Analysis (Chapter 5)	Facility Level	Risk reduction options recommended but detailed analysis of baseline risk and options needed for bulk pesticide transportation	• Quantitative risk estimate developed for baseline and options to reduce risk • Benefits of the risk reduction option input into the decision process to determine the risk reduction options to implement, if any

TRM Framework	XYZ Chemical Level of Benefit	Risk Management Issues	Risk Reduction Recommendation
Security Considerations (Chapter 6)	Facility Level	Chemical/modes escalated from corporate-level prioritization were only reviewed at the Asian facility for transportation safety risk	• Security prioritization and TSVA identified chemical/modes with escalated security concerns and developed risk reduction options • Repackaged chlorine options – Driver stops – Product stewardship – Cyber vulnerability assessment • Butadiene option – Tamper-resistant seals

7.4.1 Qualitative Risk Analysis Reduction Options

As part of the comparison activities across the XYZ Chemical operations, it was determined that there was a difference in the safety practices associated with bulk chlorine movements. The facilities in the United States and Europe had conducted benchmarking activities to determine the practices of the major chlorine producers and bulk shippers as well as the recommended practices of a chlorine trade association and Responsible Care® requirements. These practices have been implemented and are kept current in both the United States and Europe, but have not been consistently applied in XYZ Chemical's other regions of operation.

Based on this finding, it was recommended that a new corporate standard be developed and implemented for the other operating regions. Specifically, XYZ Chemical made the decision to apply the practices of the United States and Europe for the transport of bulk chlorine. These practices include carrier requirements for tank car specifications, emergency response, and routing consideration. In some cases, these practices are more stringent than the country-specific requirements for these regions.

7.4.2 Semi-Quantitative Risk Analysis Reduction Options

The results of the semi-quantitative risk analysis focused on a segment of the chlorine (ton container) route through the medium population density and the entire bulk pesticide transportation route. The recommendations developed for consideration included:

- Chlorine:
  - Consider rerouting truck deliveries through the population center as this segment of the route is not necessary for reaching the final destination.
  - Consider changing the shipment from ton containers to smaller cylinders.

- Pesticide (bulk):
  - Consider rerouting tank truck deliveries to eliminate passing near higher-population densities along the route.
  - Consider decreasing the tank truck shipment volume and increasing deliveries.

As part of the semi-quantitative risk analysis report to corporate, the analysis team included an evaluation of the recommended risk mitigation options. For chlorine, the second recommendation was not practical because the end user required ton containers; thus, the transport of smaller cylinders was not an option for the customer. Therefore, this recommendation was removed from further consideration. Since ton containers are shipped via truck, an alternate route was discussed with the carrier—a new route that eliminated passing near higher-population densities with little change in the distance or conditions of the route.

As shown in Figure 7.5, applying a revised medium/low density for this segment of the route could reduce the risk ranking for all three potential release sizes of chlorine (small, medium, and large). Implementing this recommendation will reduce the small release from a risk level of II to IV, and will reduce the risk level from II to III for the medium and large release cases. Since this recommendation reduced chlorine ton shipments from higher risk to a lower-risk level, and with minimal operational and financial impacts, this routing recommendation was approved and implemented.

For the bulk pesticide deliveries, the carrier was brought in to discuss the recommendations. The carrier was unwilling to reroute the shipments due to a number of operational and logistical issues. However, the carrier was willing to reduce the shipment size and had in its fleet designated tank trucks of half the current shipment volume. When the team reviewed this option (Option 2) with the risk analysis methodology, the same risk rankings of I and II were obtained. The results indicated that even though the shipment size was decreased, the population density along the route dictated the consequence ranking. Additionally, it was difficult to determine the effect of doubling the pesticide shipments to account for the smaller transportation volume. Therefore, for the bulk pesticide deliveries, the team recommended to corporate that this issue should be escalated for a more detailed analysis, specifically including:

- A new carrier and route
- Smaller tank trucks, but more frequent deliveries (current route) by the current carrier

Figure 7.5 Benefits of Chlorine Ton Rerouting (all release sizes)

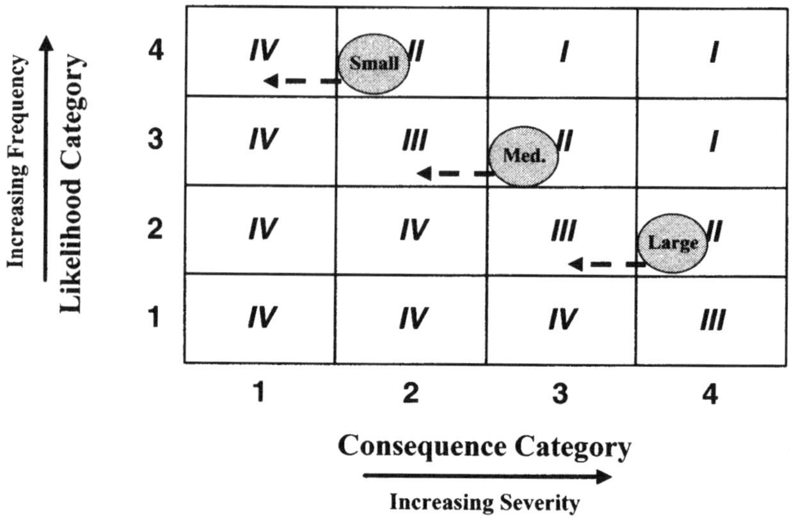

7.4.3 Quantitative Risk Analysis Reduction Options

These mitigation options—new carrier and route (Option 1) and current carrier and route with smaller tank trucks but more frequent deliveries (Option 2)—were evaluated to the current operation using the same QRA approach and assumptions. The results calculated for the current operation and mitigation options included:

- Average Individual Risk
  - Current operation: 1.9E-06 impacts per year
  - Option 1: 6.1E-07 impacts per year
  - Option 2: 1.0E-06 impacts per year

From this comparison, Option1 has the lowest average individual risk, driven by the fact that the entire route runs through a suburban population density as compared to the full range of population densities for the current operation and Option 2. Additionally, the average individual risk for Option 2 was lower than the current option due to the shorter consequence zones, which was attributable to the smaller package design (as detailed in Table 5.5).

- Societal Risk
  - Societal risk was presented in the form of an F-N curve. The F-N curve comparing the pesticide transportation options (see Figure 7.6), illustrates the following:
    - The current transportation route poses the highest risk, in terms of both maximum estimated number of impacts and the frequency of occurrence.

- Option 1 has a maximum number of impacts 67% lower than the current operation. Additionally, there was a slight reduction in the likelihood at the lower-impact levels.
- Option 2 has a maximum number of impacts 87% lower than the current operation and 60% lower than Option 1. Option 2 has the largest reduction in likelihood at the lower-impact levels.

Figure 7.6 F-N Curve Pesticide Route Option Comparison

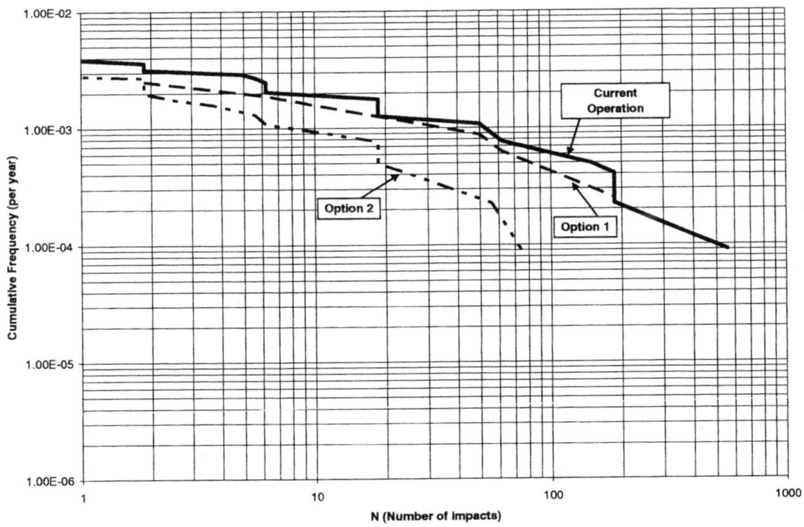

- Average Rate of Impact
  - The average rate of impact is the sum of the F-N pairs. For each option the following was also calculated:
    - Current operation: 1.8E-01 impacts/year (1 impact every 6 years)
    - Option 1: 9.4E-02 impacts/year (1 impact every 11 years)
    - Option 2: 3.2E-02 impacts/year (1 impact every 30 years)

Option 1 had the lowest individual risk, but Option 2 had the lowest societal risk and average rate of impact. Since risk criteria were not defined at the operations level in this example, the results were sent back to management for comparison to the other facilities transporting pesticide by tank truck.

Using these results, management made the following decisions regarding the transit of bulk pesticide:

- The baseline level of risk needs to be managed.
- Neither Option 1 nor Option 2 will be selected since the resulting potential did not achieve management's expected level of risk reduction.

Since this is a profitable business for XYZ Chemical, a separate analysis was commissioned to review the bulk pesticide operation and all potential risk reduction options, and combination of options that can be implemented to manage the risk of this operation. An incremental risk method (described earlier in the chapter) will be applied to identify which combination of risk reduction options provides the greatest benefit for the cost. Additionally, since XYZ Chemical's management has implemented and embraces the TRM framework, management will wait for the results of this additional study before making a final decision regarding bulk pesticide shipments. This may include a suite of risk mitigation options, or the ultimate desire of XYZ Chemical to exit the pesticide business for this facility.

7.4.4 Risk Reduction Options for Security

Based on these results the TSVA, the team recommended the following security countermeasures for consideration:

- Repackaged chlorine:
  - Consider not permitting stops during the delivery of repackaged chlorine or requiring drivers to radio in when they have stopped, report their location, and verify the security of the shipment.
  - Consider developing a formal product stewardship program for all current and future chlorine customers.
  - Consider a follow-up cyber vulnerability assessment to determine if the system is vulnerable to attack and the creation of false orders.
- Butadiene:
  - Consider requiring supplier to add tamper-resistant seals to all tank trucks, and require the facility to verify the seal before accepting it for on-site delivery.

The recommendations for bulk chlorine focused on the theft of repackaged chlorine cylinders and ton containers, and for butadiene the potential for contamination at the production site or in transit before reaching the facility. Figure 7.7 illustrates the results of these recommendations. Chlorine recommendation 1 is labeled as "Cl Transit" and the other two chlorine options labeled "Cl Cyber," which includes the combination of product stewardship and a cyber vulnerability analysis. The butadiene risk reduction option is labeled as "Butadiene Tamper," which evaluated the effectiveness of tamper-resistant seals.

As shown above, all the recommendations for security have the potential to reduce the likelihood of a security event, as follows:

- *Cl Transit* —eliminating stops during the delivery of repackaged chlorine or requiring drivers to radio in when they have stopped, report their location, and verify the security of the shipment increases the *detection* of the location of the shipments, *and deters* theft by not leaving an unattended vehicle.

- *Cl Cyber*—implementing a product stewardship program and conducting a cyber vulnerability analysis results in *detection* of customer practices that use and store chlorine, as well as identification of potentially false orders, which also *deter* improper orders.
- *Butadiene Tamper*—requiring supplier to add tamper-resistant seals to all tank trucks, and requiring the facility to verify the seal before accepting it on site will *deter* tampering in transit and enable the facility to detect tampered shipments when arriving on site.

Figure 7.7 Benefits of Security Risk Reduction Options

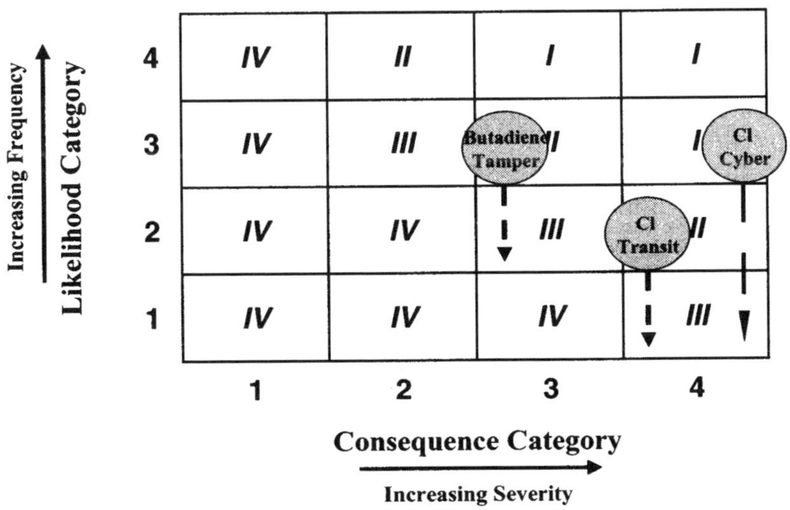

The chlorine and butadiene recommendations reduce these higher-risk operations to a risk level of III. Additionally, since there is no trade-off with safety, these recommendations, which will increase the level of security, are accepted for this facility's transportation operations for these chemicals.

REFERENCES

American Petroleum Institute, National Petrochemical & Refiners Association, *Security Vulnerability Assessment for the Petroleum and Petrochemical Industries, Second Edition*, August, 2004

Center for Chemical Process Safety, *Guidelines for Analyzing and Managing the Security Vulnerabilities of Fixed Chemical Sites*, American Institute of Chemical Engineers, New York, 2003

Center for Chemical Process Safety, *Guidelines for Chemical Process Quantitative Risk Analysis, Second Edition*, American Institute of Chemical Engineers, New York, 2000

Center for Chemical Process Safety, *Guidelines for Chemical Transportation Risk Analysis,* American Institute of Chemical Engineers, New York, 1995

Center for Chemical Process Safety, *Guidelines for Risk Based Process Safety*, American Institute of Chemical Engineers, New York, 2007

8
PROGRAM SUSTAINABILITY

As companies continue to expand globally, and supply chains continue to increase in complexity, the adaptability and sustainability of the transportation risk management (TRM) system will be a key competitive advantage. The risk management framework and safety and security analysis techniques presented in this book are flexible, adaptable, and appropriate for an uncertain environment of changing public acceptance of hazardous operations, evolving worldwide regulations, changing supply chain partners, and growing business opportunities and operational environments in developing countries. Even with the uncertainties, challenges, and opportunities that the future holds, the implementation of the TRM framework as a management system combined with the concepts, tools, and techniques presented in this book will position a company to be able to respond to changes and to manage its transportation risks in the future.

The TRM framework is not a one-time risk assessment activity, but rather builds upon the original safety management ideas process by the CCPS in the late 1980s, integrates lessons learned and operational changes, applies the management system principles of "plan, do, check, act," and organizes a system in a way that will be useful to all transportation operations—even with relatively lower hazards —throughout the life cycle of the operation. The TRM framework from Chapter 1 (Figure 1.2) fits the "plan, do, check, act" principles (Figure 8.1), as detailed below:

- *Plan*—development of a system (e.g., TRM Framework) for managing transportation risks, and includes periodic review and update of the management system, as needed
- *Do*—sustainability piece of the management system, and includes management commitment and a company culture that implements risk management concepts in daily decision-making
- *Check*—primary management systems (Chapter 2), identification and prioritization activities (Chapter 3), and additional risk analysis of elevated safety and security issues (Chapters 4–6)
- *Act*—identification, development, evaluation, and selection of risk reduction options (Chapter 7)

Figure 8.1 Plan, Do, Check, Act—TRM Framework

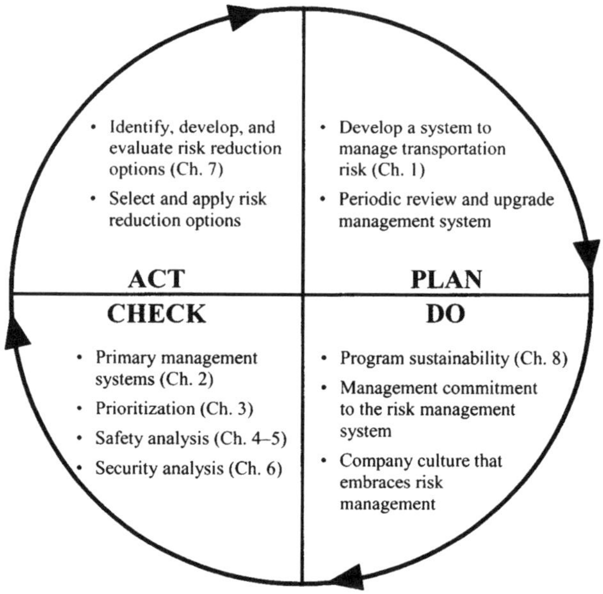

This chapter reviews the following concepts that will help ensure the sustainability of a transportation risk management program:

- Ongoing commitment to risk management
- Continuous improvement
- Emerging safety and security trends
- Evolving transportation risk analysis practices

8.1 ONGOING COMMITMENT TO RISK MANAGEMENT

One of the most critical elements of success for a transportation risk management process is the ongoing upper-level management leadership and commitment. Many of these concepts are presented in Chapter 2 as part of the development of the primary management systems. Management must show a tangible, visible commitment—including resources—to reduce and manage transportation-related risks. This visible management commitment will promote a proactive "risk reduction culture" where questions about risk are considered in making day-to-day decisions and routine operations are performed.

To promote the long-term commitment from management, the TRM system should provide a competitive advantage. This advantage will come from the ability to quickly identify and react to changes in:

- Available markets

- Product profiles
- Market leadership
- Transportation infrastructure improvements
- Access to new modes of transportation
- Safety and security technologies
- Operational technologies
- Supply chain partners and networks
- Security threats

8.2 CONTINUOUS IMPROVEMENT

The concept of continuous improvement is vital to the continued safety and security successes of the transportation industry. The goal is to improve as a result of regular, consistent efforts rather than episodic or step-wise changes, producing tangible, positive improvements either in performance, efficiency, or both. Continuous improvement efforts usually involve a formal evaluation of the status of an activity or management system, along with a comparison to an achievement goal. These evaluation and comparison activities occur much more frequently than formal audits, and the application of this concept throughout industry has resulted in the current operational practices presented in this book. Without continuous improvement, it is possible that the industry may become complacent or relax its standards. This could cause problems since the industry's existing safety and security record is the direct result of the previous and current investments. Therefore, the philosophy of continuous improvement will be the driving force for advances in safety and security practices in the future.

Additionally, continuous improvement encompasses the realization that technological advances will result in presently unavailable solutions. A risk reduction option that is either not feasible and/or too costly today may well become feasible both technically and economically in the future. Many such advancements, while having key safety and security benefits, may also result in better performance and additional competitive advantages. Examples of these types of changes may include:

- Railcar tracking that allows the supplier to know where its product is along the route can result in better customer service
- Dome and valve sealing that allows for greater confidence that materials in transit have not been compromised from a quality perspective; thus, a returned shipment would not have to be retested for QA/QC purposes, would not have to be segregated, and could more easily be redirected to another customer

The TRM framework facilitates continuous improvements through periodic assessment of system-wide risks. These activities can identify changes in the risk profile of a transportation operation (e.g., a previously "lower-risk" that has increased as a result of changes in the demographics along a route), as well as providing continuous improvement opportunities through the risk reduction strategies presented in Chapter 7.

8.3 EMERGING SAFETY AND SECURITY TRENDS

Companies will need to monitor and track emerging trends so that they can respond when these become, or are about to become, actual requirements. Most companies will not make major commitments to emerging trends, but they will build future requirements into their budget and planning cycles. These trends may include new regulatory requirements, industry guidance, and changes in public policy. As population characteristics continue to change throughout the world, and the distance between the public and the hazardous material transportation operations diminishes, society's expectations of what is an acceptable risk may change. This in turn may impact the way these materials will be handled and transported. Table 8.1 lists some trends and potential future issues that may need to be considered.

Table 8.1 Emerging Transportation Trends

Emerging Trends	Potential Issues
Regulatory Requirements	• Direction of chemical transportation safety and security regulations in developed countries • Push for global harmonization (UN labeling requirements, etc.) • Challenges and opportunities in growth geographies/developing countries • Community partnerships (communities asking/requiring industry to finance emergency response services required to manage the transportation risks of materials coming to a site) • Legislated (or potential) risk guidelines • Legislated (or potential) mitigation requirements • Best available technology requirements • Inherently safer technology requirements • Regulated background security, training, and certification requirements for transportation companies and operators • Push for special requirements for special commodities
Industry Guidance	Responsible Care®
Public Policy	• Corporate governance and corporate accountability • Toxics use reduction • Sustainability and sustainable supply chains (e.g., the need for companies to consider a reduced environmental footprint for the integrated product life cycle), including raw material supply, product manufacturing, shipment, use, and disposal
Carrier Pricing	• Material hazards • Handling requirements • Route changes/restrictions

8.4 EVOLVING TRANSPORTATION RISK ANALYSIS PRACTICES

The *Guidelines for Chemical Transportation Risk Analysis* (CCPS, 1995) provides the basis for conducting a risk analysis. This book was one of the first of its kind to take the risk analysis concepts for fixed chemical facilities and apply them to the transportation of hazardous materials. While the 1995 guidelines book lists several data sources, these data sources need to be revised and kept current.

As part of the TRM process, periodic update and review of the foundational data requirements, concepts, tools, and techniques should be considered. The transportation risk assessment process should be kept current, including:

- Incorporation of current data sources for:
  - Accident rates
  - Conditional release probabilities for specific containers
- Enhancements in methodologies and tools
- Calibration of modeling vs. actual field results
- Integration of safety, security, and environmental analyses

8.5 XYZ CHEMICAL EXAMPLE—PROGRAM SUSTAINABILITY

When evaluating risk reduction strategies for XYZ Chemical in Chapter 7, the following was determined for the bulk pesticide deliveries:

- The risk of the current transportation operation needed to be reduced.
- The two risk reduction options proposed through the quantitative risk analysis technique (new carrier and route, and current carrier with smaller trucks) did not provide the level of risk reduction management was willing to accept.

To this point in the risk analysis process, XYZ Chemical commissioned a separate study to review the bulk pesticide operation and all potential risk reduction options and combination of options that could be implemented to manage the risk of this operation. This study determined that the combination of several risk reduction options reduced the bulk pesticide risk from higher risk to lower risk. While management thought it was important to reduce identified risks, the costs of the available options were not justifiable and so the movement of bulk pesticide from this Asian facility was stopped.

Following the initial implementation of the TRM framework within XYZ Chemical, XYZ management immediately saw the benefits of this transportation management system. Over the next few years, the culture at XYZ Chemical included the evaluation of risk in all business decisions, and continued to utilize this process to manage its hazardous material transportation operations. During a periodic review of the Asian plant's transportation operations a few years later, the suite of risk reduction options recommended for bulk pesticide deliveries was reevaluated. Since the initial study, there had been several advances in tank truck design, infrastructure improvements within the country, and a considerable improvement in carrier accident rates. When XYZ Chemical reevaluated this operation in light of the current operational environment and developed a new and more cost-effective suite of risk reduction options, the benefits of managing the bulk pesticide operation now exceeded the costs.

Through the sustainability of the TRM process, XYZ Chemical continues to periodically review the overarching TRM management system, previous estimates of risk, and the appropriate suite of risk reduction options for its worldwide operations. By incorporating the current changes to the Asian operations with a new set of risk mitigation options that were not available at the time of the initial risk analysis, a once profitable business for XYZ Chemical was started back up, with the risk managed to a level approved by management.

The XYZ Chemical example illustrates the benefits of the TRM framework. It includes not only the steps to develop the management system, ensure primary management systems are current, prioritize and analyze risk, and develop and implement risk reduction options, but also the need and benefit for creating a sustainable process.

REFERENCES

Center for Chemical Process Safety, *Guidelines for Chemical Transportation Risk Analysis,* American Institute of Chemical Engineers, New York, 1995

Center for Chemical Process Safety, *Guidelines for Implementing Process Safety Management Systems*, American Institute of Chemical Engineers, New York, 1994

Center for Chemical Process Safety, *Guidelines for Risk Based Process Safety*, American Institute of Chemical Engineers, New York, 2007

INDEX

A

Accidents
  accident-initiated event, *xvii,* 27–29
  examples, 4–5
Acronyms, *xiii-xv*
Acute risk, defined, *xvii*
Adversary, defined, *xvii,* 104
Alert levels, defined, *xvii,* 104
Allision, defined, *xvii*
Asset, defined, *xvii,* 104
Asset category, defined, *xvii,* 104
Atmospheric stability, defined, *xvii*
Audits
  definition, *xviii*
  in transportation risk management system, 22–23
Average individual risk, defined, *xvii*

B

Baseline programs. *See* Primary management systems
Basic event, defined, *xviii*
Benchmarking
  definition, *xviii*
  for qualitative risk analysis, 46–48
Benefit, defined, *xviii*
BLEVE. *See* Fire and flammability hazards
Bounding group, defined, *xviii*

C

Canadian regulations, 15
Capability, defined, *xviii,* 104
Chance-of-release calculation, 78–79
Change management. *See* Management of change
Chronic risk, defined, *xviii*
Collision, defined, *xviii*
Commodity data, *xviii,* 75–76
Conditional probability
  chance-of-release calculation, 78–79
  definition, *xviii*
Consequences and consequence analysis, 29–31
  definitions, *xviii,* 27, 76
  impact, 31–32
  potential-outcome flowchart, 31
  within quantitative analysis, 80–84
  risk reduction options, 141–142
  in security vulnerability assessment, 115
  within semi-quantitative analysis, 55–61, 63–64
Continuous improvement, defined, *xviii*
Cost, defined, *xviii*
Cost-benefit analysis, defined, *xix*
Countermeasure, defined, *xix,* 104
CPQRA, defined, *xix*
Critical infrastructure, defined, *xix,* 104
Customs-Trade Partnership Against Terrorism (C-TPAT), 109–110

D

Delay, defined, *xix,* 104
Demographic data, *xix,* 76
Detection, defined, *xix,* 105
Deterrence, defined, *xix,* 105
Documentation and communication of risk-reduction strategies, 134–135

E

Effect models, defined, *xix*
Emergency preparedness and response, 19–20
  risk reduction opportunities, 144–145
European regulations, 13–14, 107
Events
  accident-initiated event, 27–29
  definitions, *xvii, xix-xxi*
  event tree analysis, 79–80
  initiating event, 28–29
  non-accident-initiated event, 27, 29
Explosion and flash-fire hazards
  in consequence analysis, 81, 82
  explosions defined, *xix*
  flash fire defined, *xx*
  vapor cloud explosion (VCE) defined, *xxiv*

F

Fire and flammability hazards. *See also* Explosion and flash-fire hazards
  BLEVE defined, *xviii*
  in consequence analysis, 81
  fireball defined, *xix-xx*
  jet fire defined, *xxi*
  pool fire defined, *xxi*
F-N curve, *xx,* 86, 88–90, 147–148
Frequency data analysis, 77–80
  definitions, *xx,* 76

G

Generic data, defined, *xx*
Glossary, *xvii-xxiv*

H

Hazards and hazard zones, *xx,* 27, 28, 80–84. *See also specific hazards*
Historical incident data, defined, *xx*

I

Identification and prioritization (safety risk assessment), 5, 25, 36–39
  cataloging hazardous materials and transportation modes, 36–37
  definition, *xx*
  example, 39–41
  flowchart, 39
  industry tools, *Apdx. B (CD)*
  interactions with other stakeholders, 37
  transportation route, sensitive areas, 37
Incidents
  definitions, *xx, xxiv,* 27
  incident outcome, 27
  reporting and investigation, 20–21
Individual risk, *xx,* 85–88
  risk contour, *xxii,* 86–88
Industry standards and guidelines, 18, 110
  Guidelines for Chemical Transportation Risk Analysis, *Apdx. A (CD)*
Intent, defined, *xx-xxi*
International Civil Aviation Organization (ICAO) regulations, 12–13
International Maritime Organization (IMO), International Maritime Dangerous Goods (IMDG) Code, 12
International regulations, 10–16
  Canadian regulations, 15
  European regulations, 13–14, 107
  Mexican regulations, 15
  U.N. Model Regulations, 10–11, 107

J

Jet fire. *See* Fire and flammability hazards

L

Layers of protection, defined, *xxi,* 105
Likelihood, 33
  definitions, *xxi,* 27
  likelihood of adversary success (L_{AS}), 105
  risk reduction options, 142–143
  within semi-quantitative analysis, 55–61, 63–64

M

Management of change, 21–22
Management system, defined, *xxi*
Meteorological data, 84
Mexican regulations, 15
Mode of transportation, defined, *xxi*

N

Non-accident-initiated event, 29
  definition, *xxi,* 27

O

Operational management, 18–19

P

Physical models, defined, *xxi*
Pipeline and Hazardous Materials Safety Administration (PHMSA). *See* United States agencies
Pool fires. *See* Fire and flammability hazards
Population data, 84–85
Prevention, defined, *xxi*
Primary management systems, 5, 9–24
  auditing, 22–23
  definition, *xxi*
  emergency preparedness and response, 19
  essential components, 16–23
  example, 23–24
  incident reporting and investigation, 20–21
  industry standards and guidelines, 18, *Apdx. A & B (CD)*
  management of change, 21–22
  operational management, 18–19
  regulatory compliance, 10–16
  structure, 10
Prioritization
  risk assessment. *See* Identification and prioritization
  security screening, 111–114
Probability, defined, *xxi*
Program sustainability, *xxi,* 7, 157–162
  example, 161–162

Q

Qualitative risk analysis, 43–54
  benchmarking. *See* Benchmarking
  checklist, 48–49; *Apdx. B (CD)*
  definition, *xxi*
  example, 50–54
  flowchart, 45
Quantitative risk analysis (QRA), 71–99
  benefits, 72
  criteria, 90–91
  data sources, 75–85
  definition, *xxii*
  example, 92–99
  flowchart, 73
  overview, 72–75
  presentation of results, 85–91
  protocol, 73–74
  selection of risk-mitigation options, 147–148
  study design, 73–74
  uncertainty, 91

R

Regulations and codes
  Canadian regulations, 15
  harmonization of regulations, 15–16
  international regulations, 11–13, 107
  Internet links, *Apdx. C (CD)*
  Mexican regulations, 15
  safety regulations, 106–110
  U.S. laws and regulations, 14–15, 107–108
Response, defined, *xxii,* 105
Risk. *See also specific risk headings*
  definitions, *xxii,* 27, 105
  formulae, 27, 115
Risk analysis, 5–6, 25
  definition, *xxii*
  evolving practices, 161
Risk assessment, 25–42
  definition, *xxii*
  identification and prioritization. *See* Identification and prioritization
  protocol, 34–36
  safety concepts, 25–27
  steps (flowchart), 26
Risk contour. *See* individual risk
Risk estimation, defined, *xxii*
Risk evaluation, *xxii,* 25
Risk index, 61–62, 85–86
Risk management
  definition, *xxii*
  ongoing commitment, 158–159
Risk matrix, 61, 62–64
Risk measures, defined, *xxii*
Risk reduction, 7, 25, 133–156
  definition, *xxii*
  documentation and communication of strategies, 134–135
  example, 148–155
  identification and evaluation, 136
  initiatives, 133–137
  opportunities, 135–136, 144–145
  options and influencing factors, 138–145
    selection of options, 145–148
Route data, *xxii,* 76

S

Safety
  emerging trends, 160–161
  risk assessment concepts, 25–27

(Safety, cont'd)
security vs. safety
risk assessment, 33–34
risk reduction, 136–137
synergies and tradeoffs, 111
Security. *See* Transportation security
Security vulnerability analysis (SVA). *See* Vulnerability
Semi-quantitative risk analysis, 43–45, 54–69, 146
benefits, 54–55
definition, *xxiii*
example, 65–69
factors to be considered, 55–61
flowchart, 45
techniques, 61–64
Sensitivity, defined, *xxiii*
Societal risk, *xxiii,* 86, 88–90
Supply chain. *See* Transportation supply chain

T

Terrorism
Customs-Trade Partnership Against Terrorism (C-TPAT), 109–110
definition, *xxiii,* 106
terrorist attacks, 101–103
Threat. *See* Vulnerability
Toxic hazards, in consequence analysis, 81, 82–84
Transect, *xxiv,* 87–88
Transloading operations, defined, *xxiv*
Transportation risk management (TRM), 3–7. *See also specific subject headings*
audits, 22–23
example, Apdx. E (CD)
framework, 5–7
key stakeholders, 2–3
Transportation safety. *See* Safety
Transportation security, 101–131. *See also* Terrorism
attacks against infrastructure, 101–103
basic concepts and definitions, *xvii-xxiv,* 103–111
cross-border initiatives, 109–110
emerging trends, 160–161
example, 124–130
industry guidance, 110
laws and regulations, 106–110, *Apdx. C (CD)*
overview, 101–103
practical recommendations and ideas, 119–123
prioritization process, 111–114
safety vs. security
risk assessment, 33–34
risk reduction, 136–137
synergies and tradeoffs, 111
security plan, 105
TSVA. *See* Vulnerability
vulnerability. *See* Vulnerability
Transportation supply chain
chemical supply chain activities, 3
definition, *xxiv*
key stakeholders, 2–3
risk reduction options, 138–140

U

Uncertainty, defined, *xxiv*
United Nations Model Regulations, 10–11, 107
United States agencies, 14–15, 108
Coast Guard, 15
DHS / TSA, 15
PHMSA / DOT, 14, 107

V

Vapor cloud explosion (VCE). *See* Explosion and flash-fire hazards
Vulnerability
definition, *xxiv,* 106
likelihood of adversary success (L_{AS}). *See* Likelihood
security vulnerability assessment, 115–119
target attractiveness (A_T), *xxiii,* 106
threats, *xxiii,* 106, 116–117
TSVA / SVA (security vulnerability analyses), 103–104
definitions, *xxiii-xxiv,* 105–106
methodology, 117–119
resources, 119

W

Worst credible incident, defined, *xxiv*